T0257992

Industrial Design Engineering

Industrial Design Engineering

Edited by **Gary Baker**

New York

Published by NY Research Press,
23 West, 55th Street, Suite 816,
New York, NY 10019, USA
www.nyresearchpress.com

Industrial Design Engineering
Edited by Gary Baker

International Standard Book Number: 978-1-63238-293-1 (Hardback)

Printed in the United States of America.

Contents

Preface

I am honored to present to you this unique book which encompasses the most up-to-date data in the field. I was extremely pleased to get this opportunity of editing the work of experts from across the globe. I have also written papers in this field and researched the various aspects revolving around the progress of the discipline. I have tried to unify my knowledge along with that of stalwarts from every corner of the world, to produce a text which not only benefits the readers but also facilitates the growth of the field.

The field of industrial design engineering is focused in this detailed book. A rapid paced altering world needs dynamic techniques and robust theories to allow designers to deal with novel product advancement landscape favorably and make a difference in an increasingly interconnected world. Designers continue stretching the limits of their discipline, and form new paths in interdisciplinary areas, consistently moving the frontiers of their practice further. This book advances on a few basic concepts, along with touching new areas of theory and practice in industrial design. It helps readers in stepping forward in their own application and in developing new design research that is appropriate and aligned with the present challenges of this fascinating field.

Finally, I would like to thank all the contributing authors for their valuable time and contributions. This book would not have been possible without their efforts. I would also like to thank my friends and family for their constant support.

Editor

Design Process Innovation

Design Thinking in Conceptual Design Processes: A Comparison Between Industrial and Engineering Design Students

Hao Jiang and Ching-Chiuan Yen

Additional information is available at the end of the chapter

1. Introduction

Design thinking is one of the most important issues in the fields of design research, as design expertise and creativity are mainly manifested through designers' cognitive processes when they are undertaking design activities, in particular during conceptual design stages [1, 2]. Majority of the design research community tend to model design thinking as a style of thinking underlining all design domains/disciplines, and complementary to scientific thinking and other non-design thinking [3-6]. Designing in fact comprises of various activities of multifaceted nature [7]. Variations of the thinking styles between different types of designers have been reported in many empirical design studies [e.g., 8 - 10]. Literature suggests that tertiary educational programs may contribute to the characteristics of thinking styles. Lawson's study indicates that design thinking may relate to learnt behaviors [7, 11]. Senior undergraduate architecture and science students demonstrated distinct problem solving strategies, while such disciplinary difference was not observed between first-year undergraduate architecture students and high school students [11]. Following the same rationale, this paper is interested in identifying possible effects of different design programs on shaping students' design thinking styles and the associated design strategies.

2. Two design programs in the National University of Singapore (NUS)

Designing is a complex human activity and encompasses a series of complex interactions between many factors or variables. Controlling some of those variables becomes necessary to provide a meaningful dataset to work with. This paper focuses on design thinking in conceptual design activities in the domain of product design. Two relevant design programs are studied, i.e. Industrial Design (ID) and Mechanical Engineering Design (ME) from the National University of Singapore (NUS), which is consistently ranked as one of the best universities in Asia and the world. These design programs themselves also respectively hold a good reputation in their own fields.

Design thinking is implemented as one of the essential pillars in these two programs. Both programs value multi-disciplinary aptitude and strongly encourage their students to take modules offered by other faculties, such as humanity, business and management, to broaden their perspectives and repertoire of skills [12, 13]. These two programs also collaborate with each other through joint design workshops and projects [14]. In addition,both programs champion a learning-by-doing approach in their curricula, while less relying on conventional lecture-based teaching and learning.

A closer examination is conducted by reviewing two capstone design courses: ID's "vertical design studio" and ME's "industry-sponsored projects". These two courses require students to undertake design projects on the basis of a small-scaled design teams (usually 2~5px). Faculty members and industry partners will co-supervise the projects. Many projects involve participation from other disciplines.

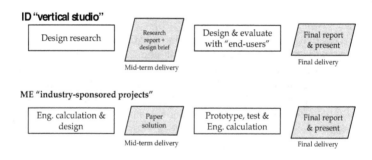

Figure 1. General schedules of ID and ME's design courses

The comparison of course plans demonstrates that the two courses have very different designing processes, as shown in Figure 1. ME projects require students to complete conceptual and embodiment design in the first half of course and submit a "paper solution" with detailed engineering calculations and drawings as their mid-term delivery. The foci of the second half of ME course is about evaluation and improvement of the proposed concept through a working prototype and further engineering calculations. ID's "vertical studio" emphasizes on the front end of designing, such as background research and problem scoping, user study, analysis of existing products, and technical and marketing inquiries. ID's mid-term delivery normally is a presentation of research findings about the problematic situation and design opportunities. The generation and development of design concepts/solutions are postponed to the second half of the course.

Another difference identified is the "given problem" presented in the task description or design briefs. ID briefs usually are in a narrative form, describing trends of design and the problematic situations. The project brief provides an abstract vision, rather than a problem. The problem statement may be presented like "to satisfy higher order of hierarchy of needs" (ID Brief A), "a user-friendly product for the doctors and patients within the digital era" (ID Brief B), "design for a reasonably foreseeable future" (ID Brief C), "to question conventional notions of luxury and challenge its relevance in the modern day context" (ID Brief D), and "to explore and create new form of objects" (ID Brief E). To respond the clarity of problems, nearly all ID briefs explicitly demand that the brief/problem needs to be evolved and continuously developed throughout the whole process. The intended solutions are open-ended in the initial briefs, e.g., "a one-off object or a collection of objects" (ID Brief A), "no fixed category" (ID Brief C), or even "cool, crazy, stunning, unbelievable" (ID Brief E).

Instead of identifying the potential concepts, ME projects' task descriptions are much more structured. They are usually formatted in a form of checklist, such as backgrounds, objects, knowledge needed, deliveries, etc. The problems presented may have already specified the type, clients and detailed parameters of the envisioned solution, such as "a swing door stopper... to auto-close a swing door panel" (ME Task A), "a cooling system using ice as thermal energy storage" (ME Task B). The design requirements are clearly described and usually measurable, such as "converting a circular motion to a linear motion" (ME Task A). Some projects require to propose an application of an existing system in a particular situation, such as "an omni-directional 'Mecanum Wheel' robotic platform with Android platform control" (ME Task C), "a robot capable of taking videos for the creation of 3D images of underground sewerage pipe" (ME Task D), or a redesign/improvement, e.g., "to improve T-Bar turning device" (ME Task E).

3. Examination of consequences of the curricular differences

The differences identified in design briefs and task descriptions suggest that the two design curricula place different emphases on the "problem finding" activities, i.e., the ways in which problems are envisaged, posed, formulated, created [15, 16]. The examination of these two curricula's influences on students' habitual design behaviors thus focuses on how ID and ME students formulate and solve design problems when given the same set of design briefs and task descriptions. The design experiment was conducted in a design-studio-like setting, as shown in Figure 2. The unit of participation was a small design team formed by 2 final-year undergraduate students. Each team was asked to perform two conceptual design tasks, one for the existing market and one for future market. A detailed description of this experiment is presented in the authors' previous paper [17].

Literature shows there are two types of problem finding, i.e., the "reactive/passive" and "proactive/purposive" problem finding [15, 18]. The former category refers to the problem recognition triggered by similarities between the current situation and a known problem type related to existing solutions/problem-solving repertories. "Purposive" problem finding refers to proactive formulation of problems which are otherwise not existed or perceived as problems. Purposive problem finding is usually claimed to be "a key aspect of creative thinking and creative performance"   [16] and perhaps more important than problem solving [19].

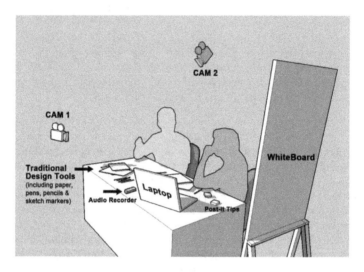

Figure 2. Experiment setup

The examination of problem finding or formulation applied Schön's reflection-in-action framework [20, 21]. Designers first "name" the relevant elements to set the boundary of problematic situation and then impose a "frame" to understand the context as well as determining actions towards solution. The analysis will focus on the quantity of named elements and how the relationships of these elements are articulated to stimulate designing processes. A small quantity of named elements and relatively direct mapping to design specifications suggest that designers do not spend much effort to proactively formulate their "own" problems to work on.

The analysis of solution development stages of designing focuses on the trajectory in which design ideas/concepts evolve. Oxman[22] proposed a multi-level structure of design knowledge from specific, context dependent precedents to more abstract, context-independent ones, as shown in Figure 3. Two distinct approaches of designing can be defined with regard to the form of the initial solution, i.e., a schema-driven refinement and a case-driven adaptation [23, 24]. The former approach starts with a highly abstract concept (i.e., a schema), and follows with a sequence of "refinement" operations to "particularize" the initial schematic state into a detailed description of a specific product [23, 24]. The latter refers to a sequence of adaptions made to transform a rather detailed concept (i.e., a case) to work in a new situation.

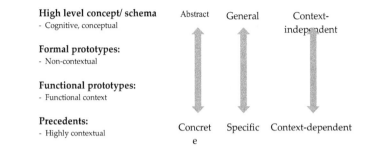

Figure 3. A structure of prior knowledge in design (adapted from [22])

Designing process transforms human needs or desires into a specification of physical embodiment. The inspiration source and form development process play an important role determining the final solution of the product. An additional analysis is then conducted to compare the "conceptual distance" between the main inspiration source and the final outcomes. This provides a coarse indicator for the design creativity.

4. Problem formulation: Framing *versus* clarification

Design problems have long been recognized as ill-defined [25], ill-structured [26] or "wicked" [27, 28]. A typical design problem usually consists of determined, undetermined and "under-determined" elements [29, 30]. The formulation of a workable design problem seems to be the first step of the designing process. Designers need to understand the problematic situation and identify the relevant considerations. Table 1 lists the elements that each team used to make sense of the initial design brief.

ID teams seemed to study the design problem from contextual points of view. They were observed spending a large amount of cognitive efforts and time to go through the problem space (e.g., potential user's profile and possible usage contexts), and to explore the potential opportunities to create something new and appropriate (cf. [17, 31]). Their problem analysis and framing activities resembled a semi-structured process, including "naming" and "framing" activities [20, 21, 30, 32]. "Mind maps" (a graphic tool to stimulate creativity and idea generation [33]) were often used to assist the discussions during this period. Figure 4 presents two examples of the graphic tools applied in early episodes of ID teams' design sessions. The main branches of graphs in the Figure show how ID teams organized their thinking process and identified the aspects of design that need to be considered. Lateral/divergent thinking [34] was demonstrated in this period which mainly aimed to enlarge the problem space. These identified elements mainly concerned about end users and potential usage contexts, rather than directly related to characteristics of design solutions.

Session\Team	1	2	3	4
ID,Task 1 (existing market)	Context People & lifestyle When What (small, personalized) How (dispense) Sensory experience (aroma)	Who Why When Where What Making process	People Type of coffee Ways to make coffee Ways to drink coffee Look Fun stuffs Market analysis Technology	People Types of coffee Ways to make coffee How to enrich experience
ID,Task 2 (future market)	Places Physical/digital Intuitive interaction Game Past/now/future Sensory experience	Context (one the go/ not on the go) Type (electric vs non-electric) Market trend Interactivity	Who What (electronic vs non-electronic) Where How (passive vs active) Sensory	People & lifestyle 5 Senses (visual, sound, smell, touch & taste) Analysis of existing products Features (existing, maybe to have)
ME,Task 1 (existing market)	Target user Type of coffee maker Size/dimension	Target user	Target user Sleek stuffs Size	Target user Size Cost
ME,Task 2 (future market)	Types of relevant products Technology (VR) Size (portability)	Types of entertainment & products	AR/VR Portable size + large display Interactive	Type of entertainment Feature (portable, easy to use)

Table 1. Elements identified to define the design problem

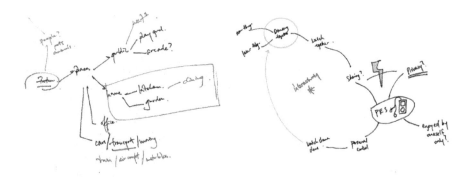

Figure 4. Mind maps used in ID team's problem framing

After identifying the key aspects required to be considered, ID teams then examined the relationships between the "named" elements and "frame" the design space in which ideation and concept development was conducted. The colored lines/texts in Figure 4 labeled designers' endeavors of connecting the identified element to formulate a coherent design "frame" that facilitate tentative design "moves" towards the solution [20, 21].

Problem analysis and framing in the ID sessions was semi-structured. Though the search of pertinent elements aimed to systematically explore the problem, it was never intended to be exhaustive. Instead, a set of particular design problem probes were used, such as human-centric factors, sensory experience, interaction, emotion, etc. (cf. Table 1). Once an promising opportunity was identified, ID teams would take "opportunistic" moves [35, 36] and propose some tentative solutions (usually in abstract forms) accordingly. There is no evidence showing ID teams may stick to a rigidly structured systematic process.

ME teams put less emphasis in understanding the design problem than ID teams did. The issues discussed in the early episodes (Table 1) were either prescribed in the design brief, such as "target users" and "types of entertainment", or related to syntactic attributes of product, such as "size/dimension" and "technology". They focused on the clarification of the problem stated in the initial task description than proactively explore the problematic situation from various perspectives. They quickly made a checklist-like "specification", which was used later to evaluate whether or not their solution fulfilled these requirements. Two examples of ME teams' named list are presented in Figure 5.

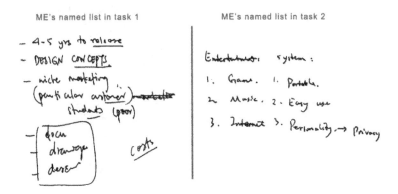

Figure 5. Specifications proposed in ME team's problem analysis

The different quantities of ID and ME teams' named elements reflect an attitudinal difference about the role of a design brief between ID and ME students. ME teams tended to treat the initial design brief as a given problem, though it may not provide a completed picture. What they ought to do is to fill the missing information and turn it into a set of measurable criteria. The stated problem was thus remain unchanged and can facilitate solution development. The "problem framing", i.e., selectively viewing a situation from various ways [20, 21], were not observed in ME sessions. ID students, on the other hand, consideredthe design problem as an imperative to innovation. The ID problem may constantly evolve when the designing process progressed. During the concept development stage, ID teams rarely made an explicit comparison between their solution and the formulated problem.

5. Solution development: "Schema-driven" *versus* "case-driven"

Preliminary exploration of design observations indicates that most design teams (both ID and ME) seldom changed their strategies when they encountered with different design tasks, though they more elaborated their solution in the design targeting at the existing market than the design for the future market. In general, design concepts were incrementally evolved from initial ideas. The difference between ID and ME's design sessions were mainly observed in the forms of their initial ideas or concepts.

As mentioned in the last section, ID teams mainly explored user and contextual issues in the early episodes of designing. The initial ideas or "primary generator" [37] proposed by them were usually highly abstract and conceptual, such as sensory experience (shown in Figure 6, left). The left panel of Figure 7 demonstrates a trajectory in which an ID concept was developed. The red arrows and annotations were labeled by the authors to visualize the flow of ideas. Many design alternatives were explored and developed in parallel. The abandoned ideas were not shown in this Figure. The keywords underlying this design were "aroma" (smell), "veil of mist" (visual) and "cute" forms. These abstract ideas were thus slowly embodied and refined through a series of thumbnails and sketches, though designers may go back and forth between different levels of abstraction.

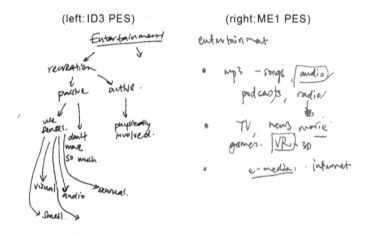

Figure 6. Two notes that assist the problem formulation process (left: ID session, right: ME session)

Development of an ID concept   Development of an ME concept

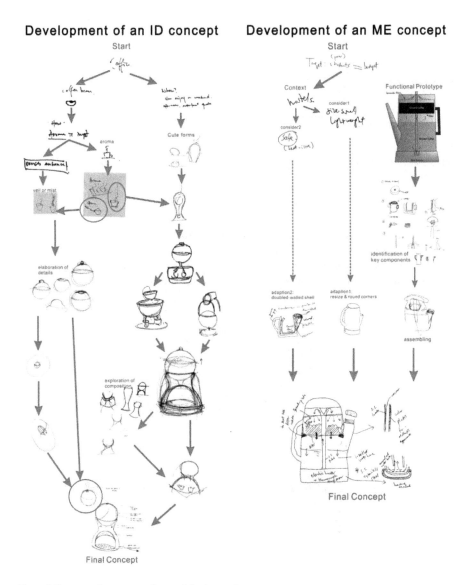

Figure 7. Two exemplar processes of concept development

During the problem analysis stage, ME teams tended to use specific precedents to understand the problematic situation. For example, in the right panel of Figure 6, a ME team used existing devices of MP3, VR, e-media, etc. to define the problem space of the design of a future entertainment device. Different from ID's "general to specific" process, ME teams' processes were usually a "specific to specific" process, i.e., adapting a rather detailed precedent or "functional prototype" to fit the current situation. This approach also refers to case-based reasoning [23]. The right panel of Figure 7 demonstrates an adaption process of designing a coffee maker on the basis of a functional prototype(i.e., a structured form of prior knowledge in design [22], cf. Figure 3). Adaptions were made with regard to considerations related to the target situation. For example, the size was scaled down to meet a single-person usage. Components, such as the shell, were modified to cut the cost. Much fewer design alternatives were observed in ME's designing processes.

In short, the solution development of ID sessions generally resembled a schema-drive refinement process, and that of ME sessions tended to follow a case-driven adaptation process [23, 24].

5.1. The inspiration sources

Figure 8 displays the main inspiration sources and the final designs from ID and ME teams. The concepts are presented in sketches and inspiration sources are represented with related photos. The association between their concepts with the inspiration sources was assisted with qualitative analysis of design processes, not just what students claimed in the concept presentations.

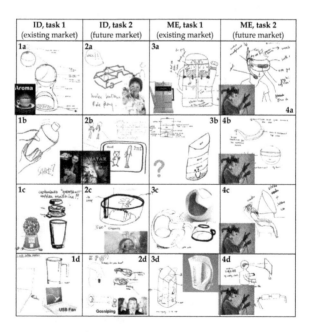

Figure 8. Design outcomes and their inspiration sources

Figure 8 shows that ID teams tended to drive their designs with a verb (e.g., experience or actions) rather than a noun [38]. For example, they proposed ideas like to "enjoy the aroma" from nice coffee (Fig. 8, 1a), to "shake" like a bartender (Fig. 8, 1b), to "refill" coffee powder like a gumball machine (Fig. 8, 1c), to "paint or draw" like a kid (Fig.8, 2a), to "share gossips" with friends (Fig.8, 2d), etc. Even when they generated forms by analogizing, it was usually undertaken in a very abstract level. As demonstrated by Figure 9, it was the feeling or emotional response that designers tried to recreate, rather than to duplicate the specific forms.

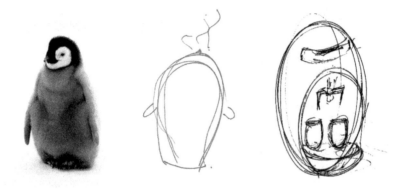

Figure 9. An inspiration source and the implements in an ID designing process

ME teams tended to build solutions on the basis of adaptation from existing products. They seemed to use product morphing or variant method [39, 40], and tended to incorporate salient features of existing products in their designs. Some incremental modifications were made, but the overall system architecture [41] were often kept untouched. Their designs were thus mainly redesigns, or "variant"/ "adaptive" designs [40]. All ME teams displayed a very high degree of similarity in their future product solutions, i.e.,a goggle-based VR system for the future entertainment design. In the design for the existing market (a coffee maker), two teams recreated a simplified version of existing products to reduce the manufacture costs. Though ME tream3 located their inspiration source outside of coffee-related products for their coffee maker concept, they almost duplicated the form of a cradle and squeezing the coffee making components into this form abruptly, as shown in Figure 10.

The above results suggest that different types of innovation strategies were preferred between these two groups of students and the preferences seemed to be independent from the nature of design tasks. ID teams seemed to be more interested on radical innovation and their designs were less attached to available design precedents. This claim echoed with Purcell and Gero's [10] study of precedence fixation effect. ME participants may mimic some characteristics of inspired sources directly and ID students shown otherwise.

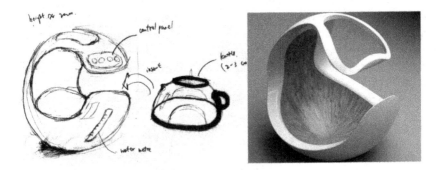

Figure 10. An ME team's coffee maker concept and its inspiration source

6. Discussions

The qualitative examination of ID and ME teams' conceptual design processes suggests that design curricula have significant impact on senior design students' habitual design behaviors. Independent of which design task they encounter with, students' designing processes generally resembled the schedule of their capstone design courses. ID teams spent considerable time and efforts to make sense of the problematic situation and purposively deferred genera- tion of solutions or partial solutions in the stage of problem framing. ME teams, on the contrary, tended to adopt a solution-oriented "problem structuring" strategy [42]. They were more likely to treat the given design brief as the mission and clarify it with envisioned solutions.

The observed behavioral differences between ID and ME students were consistent with their perceptions reported in the pretest questionnaires and follow-up interviews (cf. [43]). Though sharing similar terminologies and designing process models, ID and ME students held quite different understanding of design and designing. ME participants often held a traditional "problem-solving" view of designing. Their reports implied that the problem situations should be already prescribed in design brief, and their job is to recognize them and generate a feasible solution accordingly. ME students also tended to consider the designed product as a self- contained system, to some extent detached themselves from relevant users and contexts. On the contrary, ID students apprehended design from the perspective of its ultimate aim, i.e., "the improvement of human quality of lives" (excerpted from an ID session's conversation). The role of user (human) and usage context were usually more emphasized than that of a product per se.

Connecting these findings to the design literature, the problem formulation processes of ID and ME sessions may respectively resemble the two design paradigms, i.e., relation-in-action and problem solving [44]. ME teams mainly clarified the problem to be solved, whereas ID teams may treat it as a start point and tended to expand and reformulate the problem based on their investigation. Roozenburg and Dorst[32] argued that "problem framing" concept is proposed to challenge "technical rationality", and primarily views design as a socio-cultural construct [20, 21]. ME students, however, tend to view product as a technical/physical construct and focused on syntactic aspects of design. Problem solving model of designing perhaps is more appropriate for the technically-oriented design [30].

6.1. Teaching styles of ID and ME programs

As stated earlier, NUS ID and ME programs both champion an immersive hands-on approach of teaching and learning, while less relying on traditional lectures. Using research's terminology, the teaching styles of these two programs fall into an "inductive" approach, which learning is characterized by student-centered, active learning and collaborative learning [45, 46]. The different emphases on design problems, or varying degrees of "structuredness"/ "openness" of the problem, further distinguish ID and ME's teaching into two related but different inductive approaches, i.e., problem-based and project-based ones[45, 46]. Problem-based teaching/learning is a student-centered pedagogical approach that assumes the "centrality of problems" to learning [47]. Students work in teams to explore an open-end, ill-structure, complex (real-world) problem that usually requires knowledge from various disciplines/domains [48, 49]. A problem-based curriculum is organized around problems, and the learning process is mainly self-monitored and self-directed.

Project-based teaching/learning involves an assignment leading to the production of a final product [45, 46]. It may be subdivided into three categories according to high to low levels of student autonomy, i.e., guided project, independent project and independent inquiry [50]. The last form, independent inquiry, is overlapped with problem-based learning.

The problem statements in project-based learning are relatively well defined and the needed knowledge may be previously acquired in the past courses. ME industry projects (cf. Section 2), for example, roughly defined the scope of knowledge needed, e.g., mechanical design, heat transfer, etc. ID "vertical studio" more expects students to make speculative and exploratory propositions, such as to question the old definition of problem in the modern context or explore the vision of future. It is not about to solve a problem, but to define what the problem is. The evaluation of ID course is thus more qualitative than that of ME projects, which requires more precise and quantitative calculations.

Some researchers recommend engineering education adopting problem-based teaching/ learning approach to better prepare their students for complex real-world problems [e.g., 51]. Empirical studies have confirmed that problem-based learning has positive effects on the self-directed problem solving skills and tacit knowledge, but some negative effects are also found in the mastery of declarative knowledge [52]. Compared to speculative nature of ID, more precise requirements in engineering design more relies on a robust base of scientific knowledge. The failure of engineering design usually has severe consequences of huge cost or even human lives. Several empirical studies also show that experienced engineering designers may heavily rely on the proven solutions and make incremental refinements [53, 54]. Meanwhile, ID is much more tolerant of failure and willing to take risks. The failure of an ID concept usually has a gentler consequence than an engineering failure. This may partially account that ID studio course is more problem-based teaching/ learning whereas ME capstone course is more project-based teaching/ learning.

On the other side, the increasing demand of interdisciplinary design collaborations requires designers to better understand or appreciate their neighboring disciplines, in order to form a common ground for effective collaborations. Some fused design curricula are proposed accordingly. A design-centric engineering curriculum (DCC), for example, is recently launched by the NUS Faculty of Engineering. Many elements of design studio's approach are grafted onto engineering design courses to enhance engineering students' capabilities of "identifying and defining problems and formulating innovative and creative solutions" [55]. ID's "Design Thesis Project" module also raises the requirements on ID students' ability to develop and implement an appropriate, well-planned ID solution within the constraints of a "real world" framework of social, environmental, commercial and industrial issues, rather than simply proposing a good concept.

Author details

Hao Jiang and Ching-Chiuan Yen

Division of Industrial Design, National University of Singapore, Singapore

References

[1] Cross N. Understanding design thinking. In: Guerrini L. (ed.) Notes on doctoral research in design: Contributions from the Politecnico Di Milano. FrancoAngeli; 2011. p19-37.

[2] Cross N. Expertise in design: An overview. Design Studies 2004; 25(5): 427-41.

[3] Owen C.L. Design research: Building the knowledge base. Design Studies 1998; 19(1): 9-20.

[4] Owen C.L. Design thinking: Notes on its nature and use. Design Research Quarterly 2006; 1(2): 16-27.

[5] Archer L.B., Baynes K., Roberts P., editors. A framework for design and design education: A reader containing key papers from the 1970s and 80s. Wellesbourne: The Design and Technology Association; 2005.

[6] Cross N. Designerly ways of knowing. Basel: Birkhauser; 2008.

[7] Lawson B.R. How designers think: Demystifying the design process. 4th ed. Oxford: Elsevier/Architectural Press; 2006.

[8] Akin Ö. Variants in design cognition. In: Eastman CM, McCracken WM, Newstetter WC. (eds.) Design knowing and learning : Cognition in design education. Oxford: Elsevier Science Ltd.; 2001. p. 105-24.

[9] Akin Ö. Variants and invariants of design cognition. In: McDonnell J, Lloyd P. (eds.) About: Design: Analysing design meetings: CRC Press; 2009. p171-92.

[10] Purcell T., Gero J.S. Design and other types of fixation. Design Studies 1996; 17(4): 363-83.

[11] Lawson B.R. Cognitive strategies in architectural design. Ergonomics. 1979; 22(1): 59-68.

[12] NUS ME. Bachelor of engineering (mechanical engineering): Degree requirement. http://me.nus.edu.sg/ prospectivestudent_undergrad_req.php/ (accessed July 12, 2011).

[13] NUS DID. The synergistic three-pronged approach. 2010. http://nusdid.edu.sg/ whynusdid/ ourapproach.htm/ (accessed July 12, 1011).

[14] Fuh J.Y.H., Lu L., Quan C., Lim S.C. Product design for industry: The NUS experience. In: Proceedings of the Engineering Capstone Design Course Conference; June 2007.

[15] Getzels J.W., Csikszentmihalyi M. The creative vision: A longitudinal study of problem finding in art. New York: John Wiley & Sons Ltd; 1976.

[16] Jay E.S., Perkins D.N. Problem finding: The search for mechanism. In: Runco MA. (ed.) The creativity research handbook. Cresskill, N.J.: Hampton Press; 1997. p. 257-93.

[17] Jiang H., Yen C.C. Understanding senior design students' product conceptual design activities: A comparison between industrial and engineering design students. In: the 2010 Design Research Society (DRS) international conference "Design & Complexity"; 7-9 July 2010; Montreal, Canada.

[18] Kleindorfer P.R., Kunreuther H.C., Schoemaker. P.J.H. Problem finding and alternative generation. In: Kleindorfer PR, Kunreuther HC, Schoemaker. PJH. (eds.) Decision sciences : An integrative perspective. Cambridge: Cambridge University Press; 1993. p24-63.

[19] Runco M.A., eiditor. Problem finding, problem solving, and creativity. Norwood, N.J.: Ablex Publishing Corporation; 1994.

[20] Schön D.A. Designing as reflective conversation with the materials of a design situation. Knowledge-Based Systems 1992; 5(1): 3-14.

[21] Schön D.A. The reflective practitioner : How professionals think in action (Paperback ed.). Aldershot, U.K.: Ashgate; 1991.

[22] Oxman R.E. Prior knowledge in design: A dynamic knowledge-based model of design and creativity. Design Studies 1990; 11(1): 17-28.

[23] Ball L.J., Ormerod T.C., Morley N.J. Spontaneous analogising in engineering design: A comparative analysis of experts and novices. Design Studies 2004; 25(5): 495-508.

[24] Oxman R.E., Oxman R.M. Refinement and adaptation in design cognition. Design Studies 1992; 13(2): 117-34.

[25] Eastman C.M. Cognitive processes and ill-defined problems: A case study from design. In: Proceedings of the First International Joint Conference on Artificial Intelligence (IJCAI). Bedford, MA: MITRE; 1969.

[26] Simon H.A. The structure of ill-structured problems. In: Cross N. (ed.) Developments in design methodoligy. Chichester ; New York: John Wiley & Sons Ltd; 1984. p145-66.

[27] Buchanan R. Wicked problems in design thinking. Design Issues 1992; 8(2) :5-21.

[28] Rittel H.W.J., Webber M.M. Planning problems are wicked problems. In: Cross N. (ed.) Developments in design methodology. Chichester ; New York: John Wiley & Sons Ltd; 1984. p135-44.

[29] Dorst K. The problem of design problems. In: Cross N, Edmonds E. (eds.) Expertise in design: Design thinking research symposium (DTRS) 6; 2003. p135-47.

[30] Lawson B.R., Dorst K. Design expertise. Oxford: Architectural Press; 2009.

[31] Yen C.-C., Jiang H. Examining the dynamic processes of conceptual design: An ontologically-based protocol analysis. In: Peng Y-H, Chen C-H. (eds.) Proceedings of 2011 IDA (International Design Alliance) congress education conference. Taipei, Taiwan: Taiwan Design Center; 2011. p103-13.

[32] Roozenburg N.F.M., Dorst K. Describing design as a reflective practice: Observations on schon's theory of practice. In: Frankenberger E, Badke-Schaub P, Birkhofer H. (eds.) Designers : The key to successful product development. New York: Springer; 1998. p29-41.

[33] Buzan T., Buzan B. The mind map book: Unlock your creativity, boost your memory, change your life London: BBC Active; 2010.

[34] Goel V. Sketches of thought. Cambridge, Mass.: MIT Press; 1995.

[35] Davies S.P. Characterizing the program design activity: Neither strictly top-down nor globally opportunistic. Behaviour and Information Technology 1991; 10(3): 173-90

[36] Guindon R. Designing the design process: Exploiting opportunistic thoughts. Human-Computer Interaction 1990; 5(2): 305-44.

[37] Darke J. The primary generator and the design process. Design Studies 1979; 1(1): 36-44.

[38] Kelley T., Littman J. The art of innovation: Lessons in creativity from IDEO, america's leading design firm. London: Profile Books; 2004.

[39] Tuulenmäki A. Concepts in uncertain business environments. In: Keinonen T, Takala R. (eds.) Product concept design: A review of the conceptual design of products in industry. New York, NY: Springer; 2006. p157-75.

[40] Pahl G., Beitz W., Feldhusen J., Grote K.-H. Engineering design: A systematic approach. 3rd English ed. New York: Springer; 2007.

[41] Ulrich K.T., Eppinger S.D. Product design and development. 4th ed. Boston: McGraw-Hill/Irwin; 2008.

[42] Restrepo J., Christiaans H. Problem structuring and information access in design. Journal of Design Research 2004; 4(2): DOI: 10.1504/JDR.2004.009842

[43] Yen C.C., Jiang H. Design thinking in different disciplines. In: "The Future is" 2012 Eastman/IDSA Education Symposium, IDSA 2012 international conference; Aug 15-18; Boston 2012.

[44] Dorst K., Dijkhuis J. Comparing paradigms for describing design activity. Design Studies 1995; 16(2): 261-74.

[45] Prince M.J., Felder R.M. The many faces of inductive teaching and learning. Journal of College Science Teaching 2007; 36(5): 14-20.

[46] Prince M.J., Felder R.M. Inductive teaching and learning methods: Definitions, comparisons, and research bases. Journal of Engineering Education 2006; 95(2): 123-38.

[47] Jonassen D.H., Hung W. All problems are not equal: Implications for problem-based learning. Interdisciplinary Journal of Problem-Based Learning 2008; 2(2): 4.

[48] Kolmos A., Kuru S., Hansen H., Eskil T., Podesta L., Fink F., et al. Special interest group B5 "problem based and project oriented learning". Erasmus Thematic Network Project: TREE-(Teaching and Researc in Engineering Education) in Europe, 2007.

[49] Graaff E.d., Kolmos A. Characteristics of problem-based learning. International Journal of Engineering Education 2003; 19(5) 657-62.

[50] Lee N. Project methods as the vehicle for learning in undergraduate design education: A typology. Design Studies 2009; 30(5): 541-60.

[51] Jonassen D.H., Strobel J., Lee C.B. Everyday problem solving in engineering: Lessons for engineering educators. Journal of Engineering Education 2006; 95(2): 139-51.

[52] Dochy F., Segersb M., Bosscheb P.V.d., Gijbelsb D. Effects of problem-based learning: A meta-analysis. Learning and instruction 2003; 13(5): 533-68.

[53] Kan J.W.T., Gero J.S. Using the FBS ontology to capture semantic design information in design protocol studies. In: McDonnell J, Lloyd P. (eds.) About: Designing: Analysing design meetings: CRC Press; 2009. p213-29.

[54] Lloyd P., Scott P. Discovering the design problem. Design Studies 1994; 15(2): 125-40.

[55] NUS DCC. The design-centric engineering curriculum (DCC). http://www.eng.nus.edu.sg/ugrad/ dcc/index.html/ (accessed in 23 Sep 2011).

Chapter 2

A Bi-Directional Method for Bionic Design with Examples

Carlos A. M. Versos and Denis A. Coelho

Additional information is available at the end of the chapter

1. Introduction

Design methodologies are essential tools in the design process which provide pathways, goals and technical guidelines for the development of products. These are also critical to minimize the risks and the time of the development process of a product (Kindlein et al., 2003). It should, however, be noted that the use of a method of design for the development of the product does not in itself create or guarantee the success of a product, as this will be subject to a myriad of factors including the level of technical expertise and creativity of who makes use of the method. Previous work by Versos and Coelho (2011-a, 2011-b, 2010) and by Coelho and Versos (2011, 2010) analyzed and compared several methods for guiding bionic design that were available in literature. These analyses, in addition to providing the basis of study for the development of the methodology that is aimed in this chapter, can support designers in the selection process of the bionic design method most appropriate to the problem at hand. Previous work emphasized the necessity of integrating validation activities in bionic design processes. The development and testing of improved methods that provide greater support to designers in the pursuit of activities leading to bionic solutions is the overarching aim that this chapter seeks to contribute to satisfy.

It is up to the designer to have control and decide upon the best option and way forward (Kindlein et al., 2003) in the design process. Thus, it is necessary to specify objectives, requirements and restrictions for the design process, as well as to present and define all possible paths in order to reduce the barriers for progression and to deal with the complexity inherent to the big number of variables involved in the course of the deployment of the method. Guiding the user towards requirements satisfaction and appropriate resolution of the problem at hand is the goal of any design method, including design methods with a bionic character.

As in Nature—an environment in constant adaptation and renewal, where species evolve systematically and only the strongest survive and become adapted to the environment—the planning and development of a product must also ensure an iterative nature of the process and constant reassessment of the design process. The means available to man and his own needs and ambitions are targets of constant change. Therefore, methods for developing a product should allow for continuous adjustment and restructuring.

2. Bi-directional bionic design method proposed

To define the contours of the method of bionic design process developed and that is reported in this book chapter, two possible starting guidelines were considered: guidance in the direction from the bionic solution to the design problem and guidance in the direction from the design problem to the bionic solution. Thus, two method branches (A and B) were developed respecting each of the two alternative orientations considered for the bionic design process. The common steps in both directions of analysis (C1, C2 and C3) consist in the same activities, contain the same description and as such are applicable for the two orientations. The resulting proposition can be observed in summarized form in Tables 1 and 2, a design process starting from the design problem and oriented towards the solution (A) and a design process with orientation from the bionic solution to the design problem (B), respectively.

Steps	Description
A1 - Design brief and problem definition	- Specification of the problem to be solved by identifying the functions that it must carry out, the desired requirements and restrictions involved.
	- Preparation of the list with the schematic aspects of specification and key environmental and ecological aspects to be observed.
A2 - Reformulation of the problem	- Revision and redefinition of general problems and tasks defined in biological terms and widely applicable.
	- Questioning how Nature solves the problems or functions that are intended to be solved in the design process.
A3 - Selection of solutions	- Search for biological models and solutions that meet and solve the challenges presented through literature searches, field observations, or using open discussions with biologists and experts.
A4 - Solution analysis	- Identification and morphological analysis of structures, components, processes and functions of the biological solution, related to the problem at hand.
	- Relating the functions and requirements of the problem with the functions and features of the biological solution.
C1 - Generating concepts	- Development of ideas and concepts (in the form of sketches and 3D models) based on natural models and following the guidelines and principles obtained in the steps of analysis and definition of the biological solution and the problem.
C2 - Validation	- Verification of compliance with the requirements of the problem and validating the gains introduced by the bionic concepts developed through the validation process of the corresponding relationship between the requirements and objectives of the project to achieve the goals established.
	- Selecting the most appropriate concepts for the next step.
C3 - Detail and finish	- Making technical drawings for manufacturing, detailed descriptions of components, materials, manufacturing processes and all the considerations adequate to the type and purpose of the project.
	- Construction of prototype and presentation of results.

Table 1. Condensed description of the steps of the method of bionic design developed following the direction from the problem to the solution (A).

Steps	Description
B1 - Solution identification	- Observation of natural phenomena and identification of potential solutions or biological properties with outstanding characteristics, eligible for transfer for application to human problems.
B2 - Analysis of the solution	- Analysis and layout of a number of factors that allow perceiving the shape, structure, organization and principles of the solution. - Extraction of the fundamental principles that motivate the solution.
B3 - Reformulation of the solution	- Deduction of general principles, obtained in the previous step, in particular and in greater details and considering possible links between the biological behaviour of the solution and mechanical behaviour.
B4 - Search for a problem	- Finding, taking into account the data from the previous step, real problems, existing solutions to optimize or emerging needs that can be met with the bionic considerations identified.
B5 - Design brief and association principles	- Identification and outline of the general and specific principles for the operation of the product, the requirements and constraints of the problem and the ecological and environmental aspects to be considered for subsequent association with properties extracted from the analysis of the bionic solution.
C1 - Generating concepts	- Development of ideas and concepts (in the form of sketches and 3D models) based on natural models and following the guidelines and principles obtained in steps of analysis and definition of the biological solution and the problem.
C2 - Validation	- Verification of compliance with the requirements of the problem and validating the gains introduced by the bionic concepts developed through the validation process of the corresponding relationship between the requirements and objectives of the project to achieve the goals established. - Selecting the most appropriate concepts for the next step.
C3 - Detail and finish	- Making technical drawings for construction, detailed descriptions of components, materials, manufacturing processes and all the considerations necessary for the type and purpose of the project. - Construction of prototype and presentation of results.

Table 2. Summarized description of the stages of the bionic design method developed following the orientation from the bionic solution to the design problem (B).

Tables 1 and 2 depict the sequential organization of the methodology, although iterations are possible between the various stages of each of the two directions of analysis considered. This iteration aims to enable refinement and optimization of the design with the right steps and facilitate the analogies between the natural functions of the solution and the desired functions of the problem. In the validation phase it is possible in the methodological process to go back to any previous step. Here the aim is to be able to change, correct or improve certain aspects, taking into account the needs identified through the results of the evaluation performed in previous steps.

In the following sections the activities that are necessary for implementing the steps of both branches of the methodology developed are described.

2.1. Description of the methodology developed for the direction from the design problem to the bionic solution (A)

If the guidance for the project in question follows the direction from the identification of a design problem (a new problem or an existing one), the first task will be to draft a design brief and then defining the problem and carrying out a development process following the steps described in the following sections.

2.1.1. Step A1 – Design brief and problem definition

At this stage the problem or the human need must be specified by conducting a briefing, which should identify the function (or functions) that the project will perform as well as the actual problem and the reasons for its existence. It is also important at this stage to define the target market, i.e. who is involved with the problem and the solution, as well as the definition of where the problem is and, or, where the solution is to be applied.

For the definition of the function or functions that are intended to be carried out by the design, an auxiliary method indicated by Helms et al. (2009) is the functional decomposition of the problem or need, starting with the more complex and general function, which is subsequently decomposed into sub-functions. According to the authors, for each of these sub-functions optimization criteria can thus be defined, which are useful in further evaluation of new solutions, by measuring performance and satisfaction with the optimization criteria.

The existence of a list of requirements and restrictions, subjecting the product, is equally important in this step. Environmental and ecological variables must be included in the list and considered in routine development, production, use and final disposal of the product (Kindlein et al., 2003). Thus, these should be included in the requirements of the problem, aiming to reducing the environmental impact caused by the extraction and processing of the raw material to be used, as well as by the product production, use and the end of useful life, where issues of recycling and biodegradation must be met.

Having a clear definition of the problem, it is necessary to comprehend it in terms of Nature, i.e., translating the roles and functions of the project into sub-functions performed by natural phenomena. This step is defined as the reformulation of the problem.

2.1.2. Step A2 – Reformulation of the problem

According to Helms et al. (2009), in order to find solutions analogous to biology, designers must redefine and reshape the problems and functions in general and widely applicable biological terms, questioning, for example, "how does Nature and biological solutions do (or not do) this? ". As an example, for a function defined in the first stage as "to not suffer falls," recasting this in biological terms could mean "which features in Nature and biological solutions enable resisting, preventing and reducing lack of stability?"

The third step of the method, considering the direction of analyzing the problem towards the solution concerns the search and selection of relevant biological solutions for the design problem.

2.1.3. Step A3 – Selection of solutions

Selecting the solution models that address the nature, and, or the challenges posed, can be done through literature search or fieldwork, involving some knowledge about the habitat of samples to collect (Junior et al., 2002). It can also be done using open discussions with biologists and specialists in this field.

Some of the existing techniques, identified by Helms et al. (2009), to be taken into account in the search are the modification of the restrictions of the problem, often defined strictly and accurately, therefore reducing the search area, thereby enabling a successful search. Thus, for a problem defined as "not to suffer falls," change the restrictions into a larger search space: "stability and resistance to impact." According to those authors, in order to avoid complexity of the systems and their inherent organic nature, often demands solutions that are accessible and simple but at the same time can solve various problems at the same time. Those authors also stress the importance of this step to avoid problems of similar association and weak analogies, leading to a decline in diversity and originality of potential future design concepts based on the solution chosen. These techniques can help to meet multiple requirements that the project will respond to.

After identification of the natural system that satisfies the aims, achieves the goals or solves the problem under study, one should perform an analysis of the biological solution.

2.1.4. Step A4 – Analysis of the solution

Designers must now identify and break down the structures, components, processes and functions of the biological solution, related to the problem to solve. The issues addressed in this phase, allowing a better understanding of the functional, structural, morphological and organizational levels, can be tackled by reflecting about "what is the function?" (Junior et al., 2002). This understanding of various aspects of biological characteristics of the solution can help meet multiple requirements, including effectiveness at formal, structural, functional and organizational levels.

The functional decomposition performed in the step of defining the problem may be useful in order to relate each function or sub-function and requirement of the problem with the functions and features of the biological solution (Helms et al., 2009). Thus, the understanding of the solution will be easier. Therefore, the solution that is most relevant and feasible for the particular challenges of the project can be identified and extracted in the form of a neutral solution, which requires a maximum reduction of the structural and environmental constraints of Nature (Helms et al., 2009).

After extraction of the principles of the biological solution and according to the feasibility of implementation and the needs of the project, designers can develop ideas and concepts based on natural models, following the guidelines and principles obtained in the analysis steps of the biological solution (A4) and of the problem definition (A1). The following step is concerned with creative application of the principles and concepts generated.

2.1.5. Step C1 – Generation of concepts

For generation of ideas, designers must consider the factors that influence the effectiveness of the natural form in the solution, the factors that influence the effectiveness of the function, the effectiveness of organization or the effectiveness of communication (in accordance with the objectives of the project in question), trying to incorporate them as similarly and as faithfully as possible in the design process.

As a result of this stage, sketches and 3D models (either obtained by computer modelling and, or physical models) of the concepts developed are expected. In these concepts, besides details considering all technical and functional principles identified, analogous to the biological model, environmental aspects such as life-cycle analysis, raw material, energy and waste generated (both in the manufacturing and the life of the product), the manufacturing procedures, recyclability, reuse and biodegradation after the life of the product, and aspects of packaging and transportation thereof (Kindlein et al. 2003) should also be understood.

Moreover, in this respect Nature is assumed as the protagonist and source of inspiration, whether by requiring attention to ecological aspects of the project or by focusing on the availability of natural recyclable, reusable, renewable and biodegradable materials, which should also be considered at this stage.

As a result of the process of generating concepts one may obtain a set of alternative concepts, which perhaps are not all equally suitable as a proposed solution. In these cases it is desirable to perform an intermediate stage of evaluation of the multiple concepts, according to a structured approach, such as that proposed by Ulrich and Eppinger (2004).

After selection by formal assessment of the designed concepts it is essential to validate these against the requirements and goals set for the solution.

2.1.6. Step C2 – Validation

The validation step of this method is the process where the final concepts face the needs and requirements of the problem and where the gains brought about by bionics are assessed against a conventional solution of the project.

Accordingly, and based on the results, the information and the models obtained in the previous step enable the designer to link the specific requirements and objectives of the project with five goals to achieve (or as many as applicable) set out in this work and provided as guidelines for the corresponding validation process, shown in Table 3.

Goals to achieve	Validation process for specific purposes
Innovation of paradigm for performance features	- Conceptual analytical and illustrative images to prove the change. - The paradigm shift evidences vary depending on the type of paradigm in question (examples): - The organizational level - change from a model of centralized decision-making within the organization to a cooperative, distributed process, performed by multiple elements decision-making. - The technical level - the principle of operation, drive technology, the source of energy, among others.
Optimization of shape	- A comparative approach compared to a conventional product. Examples: - Reduction of material and weight - analysis from solid modelling. - Stability – Analysis of static centre of mass (vector mechanics). - Resistance to the maximum capacity - finite element method and test of prototypes. - Storage of objects - quantification of capacity or maximum capacity.
Satisfaction of multiple requirements	- Check objectively and as much as possible, the level that has been reached for each property implicit in each requirement. - Check if the resolution of conflicts between non-compatible properties was carried out on both sides achieving a compromise between the requirements in question.
Effectiveness of organization	- Comparison between two or more systems with the same function (including the proposed system), but with different methods of organization. - Take measured levels during effective operation (real or simulated) of systems (including the proposed system) such as execution time, energy expended, material resources, expenditures, or funds generated.
Effective communication	- Validation according to the level of communication in question: - Passive Communication (triggered by observation) - effectiveness may lie in the overlap between the meaning intended to be incorporated into the product or system by the designer and the signification readings of users or observers (empirical verification). - Active communication (process between a sender and receiver synchronously) - effectiveness evaluated from the overlap of verified posts from the transmitter to the receiver and their outcome in the receiver, which should be in accordance with what was intended by the transmitter (empirical verification)

Table 3. Aspects of validation of targets to be achieved in design processes making use of the bionic approach, with indication of specific applicable procedures.

According to the results of the validation process, there might be a need for further testing, making modifications or refinements to the models, and reassessment of the principles of the biological solution and the requirements of the problem through iterations between the steps of the method, in order to attain validation. In case of complete satisfaction, validating the results, one or more concepts can then move on to the detailing and finishing phase of the bionic design project.

2.1.7. Step C3 – Detail and finish

In the last phase of the project the considerations required for the type and purpose of project that is developing that would enable the company to place the product on the market are met. Analyses of technical, financial, environmental and market aspects are also useful for the success of a product. Technical drawings and detailed descriptions of all components of the project, descriptions of the materials used, descriptions of the process of manufacture, assembly, packaging, or instructions for use are typically conducted. It is also necessary in many cases to perform the construction of a scale prototype for display and presenting the product more realistically and assessing its feasibility. In the presentation and communication of the product, eco-marketing actions should also be considered in order to effectively convey the sustainable benefits to potential customers and consumers of the product (Camocho, 2010). The existence of monitoring activities at the end of the product development process, such as sustainability reports, checklists (eco-design checklists) that consider experiences and evaluate the product, identifying new needs, are equally relevant (Camocho, 2010).

2.2. Description of the methodology developed following the orientation from the solution to the problem (B)

Following the reverse path, the direction for the project in question from the observation of Nature and useful collection of possible solutions for future applications in projects, the first step is to identify the biological solution, progressing along the following steps shown and described below.

2.2.1. Step B1 – Identification of the bionic solution

At this stage, after the observation of natural phenomena has taken place, through aid from literature review or field research, potential solutions should be found with remarkable properties or characteristics, to be transferred for application to human problems. Subsequently, the greatest number of information concerning the identified solution is obtained to carry out the analysis of the solution.

2.2.2. Step B2 – Analysis of the solution

At this point of the design process, a number of factors is determined that enable perceiving the shape, structure, organization and functional principles of the solution. Thus, one must recognize the components or systems involved in the phenomenon under analysis, and identify the organization and morphological structure, assimilate the mechanisms, principles and levels of organization, understand how the environment influences these mechanisms, among other relevant aspects for the knowledge and analysis of the solution (Colombo, 2007).

The basic questions that must be tackled at this stage are the "why" and "how Nature works" and "what is the purpose of its form and structure" (Colombo, 2007). From this analysis, in schematic / functional notation mode, the designer can extract the principle or principles that motivate the fundamental solution.

2.2.3. Step B3 – Reformulation of the solution

The stage that follows relates to the reformulation of the solution, which aims to facilitate the search for human needs, in which the biological functions of the solution may be useful. For this purpose, with the functional principles extracted from the previous step, the designer must now deduct general and specific principles, in detail, and consider possible links between the biological and the mechanical behaviour.

After reformulation of the functional principles of the natural solution in terms of technical principles and functions, follows the search for a problem.

2.2.4. Step B4 – Search for a problem

While the search in the biological domain is restricted to a finite space of existing solutions developed by Nature, the search for a design problem can include not only some existing need but also an entirely new problem (Helms et al., 2009). The designer must thus, taking into account the data obtained during the reformulation of the solution, look for real problems that are unsolved or still have gaps, collect examples of existing solutions with the possibility of more effective and sustainable solution or identify emerging needs with yet no solutions, but that may be met with bionic considerations already identified, resulting in entirely new products. Once one has identified a potential problem related to the functional principles of biological phenomena, the next step will be drafting the design brief and its association principles.

2.2.5. Step B5 – Design brief and association principles

For a clear association between the systems and components of the biological solution and the functional aspects of the problem to be solved with bionic inspiration, this stage includes the development, identification and outline of the general and specific principles for the operation of the product. It is also essential to bring forward at this stage a list of requirements and restrictions for the product to develop, where the environmental and ecological variables are also to be included.

The fundamental objective of this step is to draw a parallel between the principles and requirements of the problem with the fundamental properties of the solution extracted from the analysis.

After understanding the analogies between the potential problem and the existing solution from the natural world, and with the aid of schematic notations, functional principles extracted from the solution and analysis of the principles and specific requirements of the problem, follows the step of developing ideas and concepts. This step is applicable in both orientations of the method.

2.2.6. Step C1 – Generation of concepts

The generation of concepts step, is common to the approach oriented from the problem to the solution and was described in section 2.15. The next phase of the method deals with the evaluation or validation of the concepts generated and is also applicable for the two orientations of the method.

2.2.7. Step C2 – Validation

According to the results of the validation process, there will be a need for further testing, modifications or refinements of the models, and reassessment of the principles of biological solution of the problem and the requirements for a new validation (see description for this step in section 2.1.6). Total satisfaction and validation of results, will enable to proceed with detailing and finishing the project.

2.2.8. Step C3 – Detail and finish

This step is common to both orientations of the developed method of analysis (see description in section 2.1.7).

It is a well-established fact that Nature is constantly learning, adapting and evolving. In a method for developing products, in particular, a bionic design process, it is beneficial to consider this teaching, making progressive drafts in successive stages of observation, problem definition, solutions analysis and validation. Thus it is important to note that even with the arrival of a drafted concept to the final stage of the method, there will always be the need to continue to improve the design and optimize the product.

2.3. Adequacy of the proposed method to support the satisfaction of the five goals focused

The genesis of the proposed method comes from a collection of methods retrieved from literature which seeks to reap the benefits of the several methods reviewed in the new combined method (and still looking as far as possible to overcome some of the shortcomings pointed out). Thus, based on subjective evaluation (and its justification) of the applicability of each of the five methods for focusing on the objectives (Versos and Coelho, 2011-a, 2011-b, 2010; Coelho and Versos, 2011, 2010), we present an analysis of the same objectives towards applicability of the proposed method.

2.3.1. Optimization of shape

Based on previous analysis (Coelho and Versos, 2011) it appears that only the method of spiral design (Biomimicry Institute, 2007) was deemed applicable to pursue this goal, with the justification for this classification attributed to the fact that it is an iterative method, explicitly, which favours systematic optimization. Since the characteristic of interaction is present in the proposed method in its two directions of analysis, it is deemed applicable to support achieving this goal.

2.3.2. Satisfaction of multiple requirements

In previous analyses, Coelho and Versos (2011) considered the Bio-inspired design method (Helms et al., 2009) as the only one of the reviewed methods applicable to support this objective (note that this is a problem-oriented approach). In the proposed method, this is considered in steps B3 (orientation from the solution to the problem) and A4 (direction from the problem to the solution), as this results from extracting from the constraints of the biological solution to make the most expeditious implementation of the principle of solution in another domain. However, these requirements are not explicitly considered after the transfer of the biological solution to the new field; considering this point, there are some shortcomings. The techniques for finding solutions, also presented in the Bio-inspired design method (Helms et al., 2009) for the selection of solutions through its various features solving several issues at the same time, also contribute to meeting this goal. In the orientation of analysis from the problem to the solution, the method developed, was considered contributing to the achievement of the satisfaction of multiple requirements in the project to be developed. The fulfilment of this goal can also be met through the consideration of environmental and ecological variables in the project, highlighted in two directions of analysis.

2.3.3. Innovation of paradigm for performance features

In previous analysis by Coelho and Versos (2011), all analyzed methods were considered applicable to provide support to achieve the objective of innovation of the paradigm for performance features. This is considered a key motivation for the proposal of each and every one of the methods previously scrutinized. It is achieved by the appearance across all the methods discussed of the processing of a biological solution so as to provide a solution to a problem inherent in a design concept. Since the proposed method considers this transformation (as in the passage from A1 to A4 and from B1 to B5) it is obvious that it satisfies this objective.

2.3.4. Effectiveness of organization

This goal was considered as fully supported through the use of the Aalborg method (oriented from the solution to the problem). The Aalborg method of analysis has achieved the category 'applicable' to achieving the goal of effectiveness of organization in view of the first stage of this method of analysis that, among other areas, focuses on the organization, structure and morphology of levels in the natural system. Given that these aspects are contemplated in both directions of analysis of the proposed method (A4 and B2), the classification of applicable is considered for this parameter for the case of the proposed bi-directional bionic design method.

2.3.5. Effective communication

Effective communication was considered a goal for which there is no support from existing methods (Coelho and Versos, 2011). Although one might consider, particularly in future work, giving support to achieve this objective, we chose not to follow this path in this work.

However, in the validation itinerary, considerations are integrated into the proposed method that are aimed at supporting the possibility of evaluating the effectiveness of communication achieved by using conventional methods to stimulate creativity. Thus, the developed method is considered applicable, albeit with gaps to be filled in the future, to support the goal of effective communication. However, for most situations, the method is applicable to support the achievement of the goal, but it cannot be achieved if it is not explicitly considered in the briefing that gives rise to the design project.

As a summary, Table 4 compares the applicability of the method developed in its two orientations of analysis, given the five key goals considered.

Goals / Direction of analysis	Optimization of shape	Satisfaction of multiple requirements	Innovation of paradigm for performance features	Effectiveness of organization	Effective communication
Orientation from the problem to the solution (A)	Applicable	Applicable	Applicable	Applicable	Applicable
Orientation from the solution to the problem (B)	Applicable	Applicable	Applicable	Applicable	Applicable with shortcomings

Table 4. Comparative analysis of the applicability of the bionic design method developed in its two orientations of analysis, given the five goals selected and considered representative of those applicable to design problems.

The development of a new methodology sought to meet the issues identified during previous study of existing methods. Steps are proposed so that the design method proposed is intended to address shortcomings in existing methods in the course of the analysis in view of their applicability to support the process to achieve five goals considered representative of the objectives pursued by those who follow a bionic approach to design. As such, we developed a descriptive method that, in addition to considering the two directions of analysis to support the validation and fulfilment of the objectives set, provides support for an iterative approach in conducting the project. It is thus meant to assist in the optimization of the results achieved with the use of a bionic approach. The method uses an approach which combines contributions of previously existing methods, which were valued by the analysis, and support of the goals listed (Versos and Coelho, 2011-a, 2011-b, 2010; Coelho and Versos, 2011, 2010). As can be seen by the comparison presented in Table 4 on the applicability of the support given to achieve the goals chosen by the proposed method (referred to in its two directions of orientation), the method supports the applicability for all combinations of goal and orientation. In addition, the proposed method achieved an increase in applicability in relation to previous methods in order to optimize and to satisfy many requirements in the orientation from the solution to the problem and for effectiveness of organization in the direction of the design process from the problem to the solution. We also considered other activities not anticipated in the methods reviewed in order to more fully support the objectives of optimizing the shape and satisfying multiple requirements. The purpose of communication effectiveness is still suffering from a lack of support for its complete satisfaction. Thus it is recommended that projects where this objective is sought, make use of other approaches described in the design literature to systematically encourage their satisfaction (e.g. Figueiredo and Coelho, 2010).

However, the proposed method while not supporting to the same degree the validation of the five objectives focused, supports validation efforts explicitly which is very distinctive of previous methods. Thus, even if not directly supporting the process leading to the satisfaction of all stated purposes, the use of this method, providing validation mechanisms, helps designers realize the level of satisfaction of each objective achieved in each iteration of the project. This assessment will assist the recognition of the need for measures to correct the detected deviations in light of the design brief objectives.

3. Examples of application of the bi-directonal method of bionic design

In order to enhance and complement the method developed, its practical application is essential. In this regard two design projects were developed that follow the method of bionic design created and presented in this chapter. This was intended not only to validate and prove its practical applicability, but also as a way to complement and in order to enhance the dissemination of the method. The deployment of the method used to guide the development of these projects is summarily shown in the following subsections, excluding steps C2 to C3. Step C1 concerns the validation stage of the method, and its content has previously been shown by Versos and Coelho (2011-a). The first design case concerns the process of design of a CD tower rack developed by the first author and following the proposed method (following the direction of analysis A—from design problem to bionic solution) with a solution inspired on the spider web. The second design example reported on, was developed starting from the elastic structures of Nature arriving at a quad-cycle with frame integrated suspension developed by the first author and following the proposed method (following direction of analysis B – from the bionic solution to the design problem).

3.1. Bionic tower for storing and holding CDs and DVDs

To consolidate and justify the method and solutions presented in this work a practical case study following the direction from the problem to the bionic solution was developed as example. The problem that has been proposed was to develop the design of a solution for bionic shelving for CDs, DVDs and books.

3.1.1. Step A1 – Problem definition

In a first step, and taking into account that the orientation of this project is initiated by identifying a problem that seeks a solution, key product requirements were established so as to define and specify the problem in question. These requirements stand in addition to the basic function of versatile storage of CDs and DVDs in their covers or books (1), in the stability against dynamic disturbances (2), and greater gripping of objects stored (3), all this as opposed to the conventional solution. Beyond these requirements other goals were still considered, specifically: increased lightness (4) against the conventional solution, ease of use through a good positioning of the spines of the CDs, DVDs and books with a view to readability (5) and a positive perception by the user in a pleasant and appealing way (6) allowing an aesthetic interest for the product to development. The last requirement, also related to communication objectives of the subject, relates to the transmission of a message an avantgarde, creative and youthful spirit (7) by the artefact. The final product should target a diverse audience in order to meet the needs and tastes of people of both genders and all ages.

Requirements (2), (4) and (5) contribute to the goal of optimizing the shape. Requirements (5), (6) and (7) contribute to the goal of effective communication. With respect to satisfying multiple requirements, in this project, this goal is achieved by the joint consideration of the objectives (1), (2), (3), (4), (5) and (6). Regarding the goal of the effectiveness of organization, this is contributed to by objective (1). The goal of paradigm innovation of performance features is a goal that does not lend binding itself to targets, and will be affected by the bionic design project results as a whole.

In order to synthesize the requirements and constraints of the problem and help the subsequent evaluation of new solutions, through satisfaction of the criteria and targets established, a Table of requirements and specifications of the problem was drawn up (Table 5).

Project Requirements	Goal to Achieve
(1) Enable storage with versatility of CDs, DVDs in their covers or books	- Effectiveness of organization
(2) Increased stability before a dynamic disturbance when compared to a conventional solution	- Shape Optimising
(3) Greater grasp of objects stored against a conventional solution	- Innovation of paradigm for performance features
(4) Increased lightness when compared to the conventional solution	- Shape Optimizing
(5) Proper positioning of the spines (CDs, DVDs and books) with a view to good readability	- Shape Optimising - Effectiveness of communication
(6) pleasant and appealing Form that allows the user to develop an aesthetic interest in the product	- Effectiveness of communication
(7) Convey a message of avant-garde, creative and youthful spirit	- Effectiveness of communication
Sustainability Requirements	**Goal to Achieve**
Reducing the environmental impact of materials: - Materials that are recyclable at the end of the product lifecycle - Biodegradable materials	
Ease of maintenance and repair	- Effectiveness of organization
Low weight of the final product's packaging for transportation	- Shape Optimising

Table 5. List of requirements and goals of the project to achieve the tower of CDs and DVDs, with details of the requirements to respect sustainable. Note that all requirements, including those of sustainable character, contribute to the goal of satisfying multiple requirements.

3.1.2. Step A2 – Reformulation of the problem

To facilitate the process of looking into the nature of biological solutions that meet the requirements of the problem, the next step was to revise the functions present in the project requirements in terms of biology and in general. Thus, a few threads of a functional nature were obtained that serve as guidance for the following step (Table 6).

Requirements	Reformulation of the requirement in terms of functions performed in Nature
Greater grasping and securing of the objects stored	-Natural solutions that capture or immobilize certain bodies Natural systems-used for the purpose of protecting organisms
Greater stability in response to dynamic disturbances	-Natural features enabling not suffering falls or impacts and resisting loads -Organisms in Nature with dimensions and morphology seemingly fragile, but with great resilience
Lightness	-Organisms, property or natural materials that are lightweight, without neglecting resistance
Reducing the environmental impact of materials	-Natural materials using renewable and biodegradable substances

Table 6. Reformulation of project requirements for CDs and DVDs tower in terms of features and functions performed by Nature.

3.1.3. Step A3 – Solution selection

Upon revising the requirements for functions and features present in Nature we sought to find, through literature review and field observations, biological solutions that best solve or respond to the topics defined. With respect to solutions that capture or immobilize natural bodies and certain natural systems used in order to protect organisms (for the requirement of greater gripping of objects) the solutions consist of cobwebs and cocoons, respectively. In addition to recognizing a similar approach between the functions held by this biological phenomenon, the cobwebs are also lighter and stronger, also responding to the requirement of lightness.

For the requirement of stability to a dynamic disturbance, taking into account the natural features that allow not suffering falls or impacts and resist efforts, organisms identified in Nature seemingly fragile, but with great resilience, were the branches of trees as an inspiring solution. Although often subject to strict conditions as the wind, and apparently fragile with modest thicknesses but reaching great length, the branches of trees show great resistance.

3.1.4. Step A4 – Analysis of the solution

The construction of the spider web, extremely lightweight and very durable - five times stronger than steel for the same cross section, can stretch more than four times its original length - aims to serve as a passive trap to "capture" insects that intersect with it, subsequently serving to feed the spider. In addition to this primary function, the cobwebs have also functionality provide support and shelter the eggs of its creator (Yahia, 2001).

The fundamental principles of the extracted biological solution (spider web) are the elastic threads, which when multiplied and combined in a particular organization, allow one to create a means of support and enclosure (principle associated with the function of gripping objects) quite sturdy and lightweight (associating the requirement for greater lightness, without loss of strength).

With respect to the second biological solution found, trees, specifically the branches, are often subject to adverse weather conditions (such as wind), bearing, despite the apparent thickness and fragility of the great lengths of their structures, high loads. This ability comes from the remarkable elasticity of the fibres in their material, tolerating movement, flexion and extension of the branches. The principle of the solution to be harvested in order to meet the requirement and functions of greater product stability is the availability of components with elastic properties which allow flexibility of the structure in case of disturbances.

3.1.5. Step C1 – Generation of concepts

After extraction of the principles of biological solutions identified the following two concepts and ideas were developed (Figures 1 and 2).

Figure 1. Sketches for the generation of the tower concept for CDS and DVDs - bionic 1.

Figure 2. Sketches for the generation of the tower concept for CDS and DVDs - bionic 2.

A system for storing CDs and DVDs in their covers structured by elastic threads, was devised, which, like the capacity of webs of silk, allows grasping objects while based on a new archetype of dematerialized ordinary shelves so as to result in a lighter solution, both visually and physically. The system of elastic threads developed visible in Figure 3 was disposed vertically and arranged - for the purpose of guiding the objects in a fixed position to organize the storage and readability of the spines - having the task of gripping objects and spatial organization.

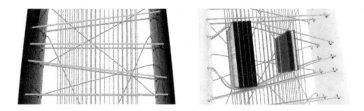

Figure 3. Guidance System of elastic threads of the towers and working sample and storage.

The first concept developed (Figure 1) for the bionic tower structure consists of a single compounding format of the tower and "webs" support system. The elastic threads are laid with pre-tensioning in holes in the structure itself, prepared for easy manual assembly or replacement of the elastics if required. A second "web" of elastic threads at the rear of tower structure was also used in order to ensure the fixing of the objects preventing them from falling. The structure of this solution was purposefully designed with a slope such as the positioning of the elastic threads horizontally and across, so as to facilitate the reading of the spines of the objects to be placed. The end result of this concept can be seen in Figure 4.

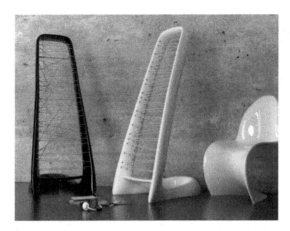

Figure 4. Representing the final appearance of bionic tower 1.

Due to the need to use some strength to place the objects in the "warp" system and because of susceptibility to dynamic disturbances in the structure of the tower, there is a need to increase the stability thereof. From the biological solution already analyzed (flexibility of the branches of trees) we developed the concept for the second tower (Figure 2). This concept, along with a light support structure (frame) of the "web" system, has a separate base connected to the frame via a third element consisting of a material with elastic properties, to ensure the flexibility of the structure. In case of a disturbance, it will provide greater stability to the tower, which will behave just as a branch of a tree, and (Fig. 5).

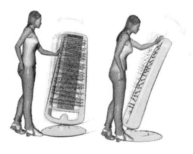

Figure 5. Demonstration of the functionality of the joint at the base of tower 2, responding to user interaction movement, increasing the bionic nature of the solution for its dynamism.

For this resilient member, its function is to store strain energy when the user intentional tilts the tower (Fig. 5) so that it may be positioned in slope for the convenience of the user and so that the insertion force for a new object in the tower is reflected in the same inclination temporarily. The elasticity of this element is given by the properties of the elastomer selected from natural origins (similar to that used in elastic filaments selected for the concept presented in the following section). To create additional rigidity a component arranged vertically inside this element is incorporated, considering the feasibility of using a bamboo cane segment. This process of selection and sizing, beyond the scope of the work presented here, will bring the possibility of gauging the dimensions of the components of this element and its proportion in future tests, with a full-scale working prototype.

Apart from this enhanced property, it is also possible to introduce sand or water inside the tower base, in order to increase the mass of this element, contributing to the enhancement of stability (displacement of the centre of mass of the tower) without compromising the size and low weight of the tower in transportation or distribution stages of the product.

Figure 6 represents the final look of the bionic tower 2, with different chromatic versions.

Figure 6. Representing the final appearance and colour studies for bionic tower 2.

3.2. Optimization of structures according to the rules of nature

Through observations of Nature and literature searches, it is possible to gather information about possible biological solutions useful for application in projects. This is based on the analysis of natural structures secrets for lightness, durability and resistance to various efforts or conditions and finding ways to link new solutions to existing or novel human needs to the solutions and rules of Nature.

3.2.1. Step B1 – Solution identification

In Nature there are numerous structures that grow and develop in order to adapt to the conditions to which they are subject. In examples present either in flora and fauna, there are small structural solutions which represent much of the secret to optimize resistance. In trees, for example one may find technical solutions to increase the resistance of structures and efforts to prevent fragility. Other structures such as bones or skeletons are also examples of inspiration because they demonstrate being as light as possible and at the same time as strong and resistant as required. in the analysis of the solutions identified, the book "Secret Design Rules of Nature", by Mattheck (2007) was considered as a reference.

3.2.2. Step B2 – Analysis of the solution

According to Mattheck (2007), observing the teachings of the structures and growth of trees can solve various problems related to efficiency of formal structures, or to eliminate or reduce the presence of cracks caused by accumulations of stresses (the reason for the ruin of a structure). According to the author, trees develop in order to strengthen the weaker areas of their structures. One of the main examples is how the base of the trunks develop in order to sustain the tree and the stresses to which it is subjected. The trunks develop more zones of connection to the ground, especially in directions exposed to the wind in order to reinforce the regions of greatest tension and avoid cracks (Mattheck, 2007). The branches and trunks of smaller cross-sections also demonstrate great resistance to impacts and dynamic forces through their flexible and elastic properties. The principle used here, as in all organisms is Nature, is of adaptation to the environment and surrounding conditions. In contrast, man-made mechanisms tend to resist and counteract adversity.

The same author discloses a method (tensile triangles) based on the growth of tree structures, to reduce the accumulation of stresses at weak points (cracks susceptible to) and to counteract potential sites of fracture. The method consists in the introduction of a square triangle, with the two acute angles of 45 degrees, symmetrically about the corner of the structure. The introduction of this triangle creates new fragile zones, susceptible to cracks, though less dangerous than the initial one. A new triangle is inserted symmetrically to the corners of the triangle resulting from the first approach, and so on to reduce the angle of fragile zones. According to the author, usually three triangles in the desired direction are sufficient. This method is similar to what the CAO (Computer Aided Optimization) method performs. The representation of this optimization system is observable in many natural structures, as for example in trees. This method can also be used in order to eliminate unused areas and components subject to excessive stresses, reducing the cross-sectional area of the component and providing optimal distribution of the stresses to a minimum area, avoiding wastage of material and reducing the weight and volume of structures. These features of formal optimization are also observed in the structures of bones or skeletons and are applied by the SKO (soft kill option) method.

3.2.3. Step B3 – Reformulation of the solution

Trees are elastic structures; such elasticity is desirable in artefacts used by man, especially in structures that are subject to dynamic loading. The solutions of optimization and adaptation to the environment that natural bodies present may contribute to the structural effectiveness of a product in order to adapt this same structure, subjected to stresses and dynamic conditions, with better results. From another perspective, the optimization and formal characteristics observed in skeletal bones (mild, simple and resistant) may contribute to the reduction of the cross-sectional overall area of a structure and hence to its dimensions and its weight without compromising strength and resistance to the loads endured.

3.2.4. Step B4 – Problem searching

Powered vehicles and propulsion for human endeavours necessarily imply a lightweight and practical manner to enable and facilitate human activity. For sites with flat terrain and the difficulty of rapid movement of these vehicles is reduced, justifying the existence of many bicycles, skates, skateboards, scooters and even tricycles, among others. In most rugged and mountainous terrain rolling on these vehicles is difficult, making it uncomfortable for the user and also causing damage to structures that are not prepared for the conditions. An existing solution, present in mountain bikes, is strengthening and equipping the structure with mechanical suspension components that allow control of the oscillations. Such a solution increases the costs of the vehicle and does not solve in a complete manner other associated problems.

For those who need to perform usual activities in hilly terrain, as evidenced in activities such as monitoring forests, difficulties are felt related to the actual movement or transport of other equipment necessary for the practice of the actual tasks. The bicycle is a vehicle more accessible for these functions but is revealed impractical in most scenarios meandering lack of stability, because of having only two wheels. The lack of comfort in travelling is another of these problems.

3.2.5. Step B5 – Design brief and association principles

According to the needs and problems identified in the previous step, requirements were established to guide the design of a vehicle powered by human effort in mountainous terrain. Thus, as a first requirement, it is desired to design a vehicle that allows access to rough roads (1) with improved comfort and stability when compared to a bicycle. This requirement is joined by the design of a damping system (2) of oscillations and impacts without use of a mechanical components suspension. The vehicle use should be sufficiently mild so as to reduce the effort required to operate and to facilitate human activity (3). The final product should also allow adaptation of fittings (4) for transport of tools to support the development of other activity by the user.

For the goal of optimizing the shape of the required object contribute to requirements (1) and (3), while the latter requirement (4) goes against the goal of organizational effectiveness. In what concerns innovation of the paradigm inherent to the performance of functions, requirement (2) directly contributes to this and is to be achieved with the proposed bionic design. The joint consideration of all requirements aims satisfying multiple requirements.

In order to establish a relation between the conditions and the properties of the problem, in analyzing the extracted fundamental solution, Table 7 was prepared, an association between the early solution and the extracted requirements inherent the problem and the identification of the environmental and ecological variables desired.

Principles derived from the solution	Project requirements
-Structures with elastic properties, as tree structures' analogy -Solutions with formal optimization and adaptation to the environment where trees are	(1) Access to rustic trails with greater comfort and stability
-Structures with elastic properties, as analogy to the structures of the trees	(2) System for damping of oscillations and impacts, without the use of mechanical suspension components
-Optimization of form through formal observation of forms and properties of skeletons and bones (light and strong)	(3) Lightness of product to reduce the effort required to operate it
	(4) Enable fitting of accessories for transportation
Environmental and ecological variables	
Reducing the environmental impact of materials:	
Materials that are recyclable at the end of product life cycle	
Biodegradable materials	

Table 7. Association between the principles extracted from the solution and the project requirements, identifying the environmental and ecological aspects to be respected.

3.2.6. Step C1 – Generating concepts

In the generating concepts phase, creative ideas and principles extracted from the solution and the project requirements, associated in the previous step, were developed and considered. Thus, the development of a structural form was sought based on the solutions and methods shown by Mattheck (2007) and in accordance with the teachings of the structures of trees and skeletons.

In order to provide greater stability and effectiveness in hilly and rough terrain, a four-wheeled vehicle for one person (see Figure 7), designated Biocross was conceived through a single structure and continuous lines, resembling the skeleton of an animal. The design of this object followed the steps in the methodology proposed in this chapter (steps B and C).

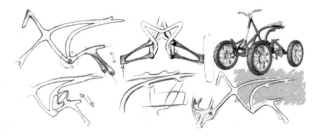

Figure 7. Sketches illustrating concepts generated for a bionic pedals vehicle and representation of a perspective drawn from the concept developed.

Figure 8 shows the method bionic optimization of structures using the approach proposed by triangles voltage Mattheck (2007), adopted in the design of the vehicle structure, which aims to reduce and eliminate critical areas of accumulation of tensions, providing Optimum reduction material.

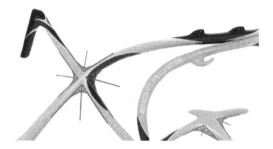

Figure 8. Formal optimization method used in the design of the bionic pedal vehicle structure.

The four-wheel connection structure itself is tapered to allow zones of greater flexibility, thus dampening the oscillations caused by terrain (Figure 9). The purpose here, as in trees or any body of Nature, is to allow the structure to adapt to environmental conditions to which it is subjected, unlike conventional mechanical components (damper and spring) that seek to counteract and withstand the irregularities of the terrain.

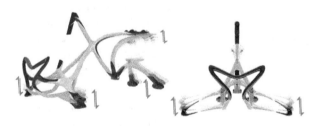

Figure 9. Representation of the effect provided by the suspension of the bionic pedals vehicle.

In order to ensure greater comfort and convenience to the user, a seat was built larger than those of ordinary bicycles. At the border of the front structure of the vehicle protrudes a protective guard thereto. In order to meet the requirement which calls for adjustment accessories for carrying utensils, at the rear of the structure itself, support bases and hooks for fastening accessories were added. All other components of the vehicle were considered standard, rendered through CAD modelling conducted for the purpose of displaying the full product (Figure 10) and perform solid mass tests.

Figure 10. Complete visual representation of the vehicle bionic developed with two chromatic versions of the Biocross bionic pedals vehicle.

Responding to the desired requirement that calls for reducing the environmental impact of the materials used bio-polymer (PLA) Ingeo Biopolymer 3251 was selected, which is also used in the design of the first bionic project described in this chapter.

4. Concluding remarks

The need for evaluation and validation during the development of bionic design projects enabling to measure product success in meeting efficiency targets and proposed requirements, was one of the evident missing features of previously existing methodologies for bionic design and which were met with the proposed methodology. Besides these aspects, the proposed method is also intended to support an iterative approach in conducting design projects in order to achieve optimal results and correct the detected deviations meeting the proposed objectives and needs. Implementation of the proposed method in practice aims its validation and also confirmation of the gains introduced in projects that follow the methodology for the process of design with inspiration taken from Nature. This is explicit from the results obtained during the two projects, which in addition to validate the method, serve as a complement to present the method.

Bionic design, a discipline capable of enriching projects with gains in efficiency, aesthetics and sustainability and with a wide margin for improvement and with a whole world where inspiration can be reaped from, will certainly bring benefits to designers in the future development of their concepts and their research. One of the studies included in this theme that could be accomplished in the future, with the objective of its development and expansion, would be an empirical study made by surveying designers in businesses that would allow identifying the actors who make use of the bionic methodology, principles and approach in everyday professional life. With the same objective, the application of a methodology for comparative analysis of the gains brought by bionics to a wide range of products would be equally interesting. It is also important to note that the method presented in this work, like any other, is not considered perfect or timeless. The evolution of scientific and biological knowledge, emerging technologies and the principles of sustainability provide new insights and new creative processes and designs. The design method should therefore be seen as a process of constant improvement, optimization and evolution – as in Nature.

Acknowledgement

The research presented in this chapter was developed as part of the first author's Master of Science thesis in industrial design engineering and as part of his ongoing doctoral studies, both supervised by the second author. A selection of results from the projects reported in this chapter have previously appeared in the conference papers Coelho & Versos (2010) and Versos & Coelho (2010), as well as in a peer-reviewed journal paper by Coelho & Versos (2011) published by Inderscience and in a peer-reviewed book chapter and a peer-reviewed journal paper by Versos & Coelho (2011-a, 2011-b) published by InTech and Common Ground, respectively.

Author details

Carlos A. M. Versos and Denis A. Coelho

Universidade da Beira Interior, Portugal

References

[1] Biomimicry Institute (2007) 'Biomimicry: a tool for innovation', available at http://www.biomimicryinstitute.org/about-us/biomimicry-a-tool-for-innovation.html (accessed on 29 December 2009).

[2] Camocho, D. (2010) 'Design Cerâmico—Guia de apoio ao Desenvolvimento Sustentável do Sector' Dissertação de Mestrado em Design e Cultura Visual, ramo de especialização em Design de Produção Industrial. Instituto de Artes Visuais, Design e Marketing, Portugal. [in Portuguese—'Ceramic Design—Guide to support the Sustainable Development Sector' Master's Thesis in Design and Visual Culture, specialization in Industrial Design Production. Institute of Visual Arts, Design and Marketing].

[3] Coelho, D. A., Versos, C. A. M. (2011) A comparative analysis of six bionic design methods, International Journal of Design Engineering 4 (2), 114-131.

[4] Coelho, D. A., Versos, C. A. M. (2010) An approach to validation of technological industrial design concepts with a bionic character, Proceedings of the International Conference on Design and Product Development (ICDPD'10), Athens, Greece, 40-45.

[5] Colombo, B. (2007). 'Biomimetic design for new technological developments' in Sal-mi, E., Stebbing, P., Burden G., Anusionwu, L. (Eds) Cumulus Working Papers, Helsinki, Finland: University of Art and Design Helsinki, pp. 29-36.

[6] Figueiredo, J. D., Coelho, D. A. (2010) Semiotic analysis in perspective: a frame of reference to inform industrial design practice. Design Principles and Practice: An International Journal. Volume 4, Issue 1, pp. 333-346.

[7] Helms, M., Vattam, S.S., Goel, A. (2009) 'Biologically inspired design: process and products', Design Studies, Vol. 30, No. -, pp. 606-622.

[8] Junior, W., Guanabara, A., Silva, E., Platcheck, E. (2002) 'Proposta de uma Metodologia para o Desenvolvimento de Produtos Baseados no Estudo da Biónica'[in Portuguese—'Proposal for a Methodology for Product Development Based on the Study of Bionics'], Brasília: P&D—Pesquisa e Design..

[9] Kindlein, W.J., Cândido, L. H. A., Platcheck, E., (2003) 'Analogia entre as metodologias de desenvolvimento de produto atuais, incluindo a proposta de uma metodologia com ênfase no ecodesign', Congresso Internacional de Pesquisa em Design, Outubro 2003 ['Analogy between the methodologies of product development today, including the proposed methodology with an emphasis on eco-design' International Conference on Design Research, October 2003], Rio de Janeiro: Anped.

[10] Mattheck, C. (2007) 'Secret Design Rules of Nature', Verlag Forschungszentrum Karlsruhe.

[11] Ulrich, K.T., Eppinger, S.T., (2004) Product Design and Development, international edition, McGraw-Hill.

[12] Versos, C. A. M., Coelho, D. A. (2011-a) Biologically Inspired Design: Methods and Validation, in Industrial Design—New Frontiers (edited by Denis A. Coelho), Intech, 101-120.

[13] Versos, C. A. M., Coelho, D. A. (2011-b) An Approach to Validation of Industrial Design Concepts Inspired by Nature, Design Principles and Practices: an International Journal 5 (3), 535-552.

[14] Versos, C. A. M., Coelho, D. A. (2010) Iterative Design of a Novel Bionic CD Storage Shelf Demonstrating an Approach to Validation of Bionic Industrial Design Engineering Concepts, Proceedings of the International Conference on Design and Product Development (ICDPD'10), Athens, Greece, 46-51.

Chapter 3

TRIZ: Design Problem Solving with Systematic Innovation

Helena V. G. Navas

Additional information is available at the end of the chapter

1. Introduction

Systematic innovation is crucial for increasing design effectiveness, enhancing competitiveness and profitability. Enterprises need to invest in systematic innovation if they want to win or survive. Innovation can no longer be seen as the product of occasional inspiration. Innovation has to be learned and managed. Innovation has to be transformed into a capacity, not a gift.

Unexpected occurrences, inconsistencies, process requirements, changes in the market and industry, demographic change, changes in perception or new knowledge can give rise to innovation opportunities.

Systematic innovation can be understood as a concept that includes the inventive instruments necessary to invent the right things that need to be invented, and incorporate them into new products and processes. Design activity needs innovation with the right dose and in the nick of time.

The Theory of Inventive Problem Solving (TRIZ), Brainstorming, Collateral Thinking, Mind Maps and other methodologies can stimulate individual and collective creativity.

2. Theory of Inventive Problem Solving (TRIZ)

The Theory of Inventive Problem Solving, better known by its acronym TRIZ was developed by Genrich Altshuller, from 1946 [1]. TRIZ is a theory that can help any engineer invent.

The TRIZ methodology can be seen and used on several levels. The highest level, the TRIZ can be seen as a science, as a philosophy or a way to be in life (a creative mode and a permanent search of continuous improvement). The more practical level, the TRIZ can be seen as a set of analytical tools that assist in the detection of contradictions on systems, in formulating and solving of design problems through the elimination or mitigation of contradictions [2].

The TRIZ methodology is based on the following grounds:

• Technical systems.

• Levels of innovation.

• Law of ideality.

• Contradictions.

Every system that performs a technical function is a technical system. Any technical system can contain one or more subsystems. The hierarchy of technical systems can be complex with many interactions. When a technical system produces harmful or inadequate effects, the system needs to be improved. Technical systems emerge; ripen to maturity, and die (they are replaced with new technical systems).

Altshuller's analysis of a large number of patents reveals that inventive value of different inventions is not equal. Altshuller systematized the solutions described in patent applications dividing them into five levels [3]:

• Level 1: routine solutions using methods well known in their area of specialty. The Level 1 is not really innovative. This category is about 30% of the total.

• Level 2: small corrections in existing systems using methods known in the industry. About 45% of the total.

• Level 3: major improvements that solve contradictions in typical systems of a particular branch of industry. About 20% of the total. This is where creative design solutions appear.

• Level 4: solutions based on application of new scientific principles. It solves the problem by replacing the original technology with a new technology. About 4% of the total.

• Level 5: innovative solutions based on scientific discoveries not previously explored. Less than 1% of the total.

The TRIZ aims to assist the development of design tasks at levels 3 and 4 (about a quarter of the total), where the simple application of traditional engineering techniques does not produce notable results.

The Law of Ideality states that any technical system tends to reduce costs, to reduce energy wastes, to reduce space and dimensional requirements, to become more effective, more reliable, and simpler. Any technical system, during its lifetime, tends to become more ideal.

We can evaluate an inventive level of a technical system by its degree of Ideality.

There are several ways to increase an ideality of a technical system.

The TRIZ axiom of evolution reveals that, during the evolution of a technical system, improvement of any part of that system can lead to conflict with another part.

A system conflict or contradiction occurs when the improvement of certain attributes results in the deterioration of others. The typical conflicts are: reliability/complexity; productivity/precision; strength/ductility, etc.

Traditional engineering and design practices can become insufficient and inefficient for the implementation of new scientific principles or for radical improvements of existing systems. Traditional way of technical and design contradictions' solving is through search of possible compromise between contradicting factors, whereas the Theory of Inventive Problem Solving (TRIZ) aims to remove contradictions and to remove compromises.

The inconsistencies are eliminated by modification of the entire system or by modification of one or more subsystems. TRIZ systematizes solutions that can be used for different technical fields and activities.

In TRIZ, the problems are divided into local and global problems [1]. The problem is considered as local when it can be mitigated or eliminated by modifying of a subsystem, keeping the remaining unchanged. The problem is classified as global when it can be solved only by the development of a new system based on a different principle of operation.

Over the past decades, TRIZ has developed into a set of different practical tools that can be used together or apart for technical problem solving and design failure analysis.

Generally, the TRIZ's problem solving process is to define a specific problem, formalize it, identify the contradictions, find examples of how others have solved the contradiction or utilized the principles, and finally, apply those general solutions to the particular problem.

Figure 1 shows the steps of the TRIZ's problem solving.

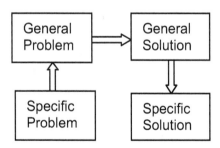

Figure 1. Steps of the TRIZ's algorithm for problem solving [4]

It is important to identify and to understand the contradiction that is causing the problem as soon as possible. TRIZ can help to identify contradictions and to formalize problems to be solved. The identification and the formalizing of problems is one of the most important and difficult tasks, with numerous impediments. The situation is often obscured.

The problem can be generalized by selecting one of the TRIZ problem solving tools. The generic solutions available within TRIZ can be of great benefit at choosing of corrective actions.

The integral development of TRIZ consists of a set of concepts [5]:

- Problem formulation system.

- Physical and technical contradictions solving.

- Concept of the ideal state of a design.

- Analysis "Substance-Field".

- Algorithm of Inventive Problem Solving (ARIZ).

Altshuller found that, despite the great technological diversity, there is only 1250 typical system conflicts. He also identified 39 engineering parameters or product attributes that engineers usually try to improve.

Table 1 presents the list of these parameters.

1. Weight of moving object	21. Power
2. Weight of nonmoving object	22. Waste of energy
3. Length of moving object	23. Waste of substance
4. Length of nonmoving object	24. Loss of information
5. Area of moving object	25. Waste of time
6. Area of nonmoving object	26. Amount of substance
7. Volume of moving object	27. Reliability
8. Volume of nonmoving object	28. Accuracy of measurement
9. Speed	29. Accuracy of manufacturing
10. Force	30. Harmful factors acting on object
11. Tension, pressure	31. Harmful side effects
12. Shape	32. Manufacturability
13. Stability of object	33. Convenience of use
14. Strength	34. Repairability
15. Durability of moving object	35. Adaptability
16. Durability of nonmoving object	36. Complexity of device
17. Temperature	37. Complexity of control
18. Brightness	38. Level of automation
19. Energy spent by moving object	39. Productivity
20. Energy spent by nonmoving object	

Table 1. Engineering parameters according to TRIZ [3]

All of these 1250 conflicts can be solved through the application of only 40 principles of invention [3], often called Techniques for Overcoming System Conflicts, which represent the Table 2.

1. Segmentation	21. Rushing through
2. Extraction	22. Convert harm into benefit
3. Local quality	23. Feedback
4. Asymmetry	24. Mediator
5. Combining	25. Self-service
6. Universality	26. Copying
7. Nesting	27. Inexpensive, short-lived object for expensive,
8. Counterweight	durable one
9. Prior counter-action	28. Replacement of a mechanical system
10. Prior action	29. Pneumatic or hydraulic construction
11. Cushion in advance	30. Flexible membranes or thin film
12. Equipotentiality	31. Use of porous material
13. Inversion	32. Changing the color
14. Spheroidality	33. Homogeneity
15. Dynamicity	34. Rejecting and regenerating parts
16. Partial or overdone action	35. Transformation of the physical and chemical states
17. Moving to a new dimension	of an object
18. Mechanical vibration	36. Phase transformation
19. Periodic action	37. Thermal expansion
20. Continuity of a useful action	38. Use strong oxidizers
	39. Inert environment
	40. Composite materials

Table 2. Invention principles of TRIZ

However, most of the principles of invention of Table 2 have a specific technical meaning introduced by Altshuller. For example, the principle of Local Quality [6]:

- Transition from a homogeneous structure of an object or outside environment/action to a heterogeneous structure.

- Have different parts of the object carry out different functions.

- Place each part of the object under conditions most favourable for its operation.

Altshuller built a contradictions matrix, classifying them as follows [1]:

- Physical Contradiction - occurs when two mutually incompatible requirements refer to the same element of the system.

- Technical Contradiction - occurs when the improvement of a particular attribute or characteristic of the system causes the deterioration of another attribute.

The first step in the conflict solving process is drawing up a statement of the problem in order to reveal the contradictions contained in the system. Then, the parameters that affect and improve system performance are identified.

The rows of the table of contradictions are then populated with parameters whose adjustment improves the behavior of the system, and these intersect the columns with parameters whose adjustment produces unwanted results. At the intersection are the numbers of invention principles that are suggested as being capable of solving the contradiction (Table 3).

In the Table 3, the rows and columns refer to the Table 1. The numbers in cells refer to the Table 2.

Characteristics		Characteristic that is getting worse												
		1	2	3	4	5	6	7	8	9	10	11	12	13
1	Weight of a mobile object	■	-	15, 8 / 29, 34	-	29, 17 / 38, 34	-	29, 2 / 40, 28	-	2, 8 / 15, 38	8, 10 / 18, 37	10, 36 / 37, 40	10, 14 / 35, 40	1, 35 / 19, 39
2	Weight of a stationary object	-	■	-	10, 1 / 29, 35	-	35, 30 / 13, 2	-	5, 35 / 14, 2	-	8, 10 / 19, 35	13, 29 / 10, 18	13, 10 / 29, 14	26, 39 / 1, 40
3	Length of a mobile object	8, 15 / 29, 34	-	■	-	15, 17 / 4	-	7, 17 / 4, 35	-	13, 4 / 8	17, 10 / 4	1, 8 / 35	1, 8 / 10, 29	1, 8 / 15, 34
4	Length of a stationary object	-	35, 28 / 40, 29	-	■	-	17, 7 / 10, 40	-	35, 8 / 2, 14	-	28, 10	1, 14 / 35	13, 14 / 15, 7	39, 37 / 35
5	Area of a mobile object	2, 17 / 29, 4	-	14, 15 / 18, 4	-	■	-	7, 14 / 17, 4	-	29, 30 / 4, 34	19, 30 / 35, 2	10, 15 / 36, 28	5, 34 / 29, 4	11, 2 / 13, 39
6	Area of a stationary object	-	30, 2 / 14, 18	-	26, 7 / 9, 39	-	■	-	-	-	1, 18 / 35, 36	10, 15 / 36, 37	-	2, 38
7	Volume of a mobile object	2, 26 / 29, 40	-	1, 7 / 4, 35	-	1, 7 / 4, 17	-	■	-	29, 4 / 38, 34	15, 35 / 36, 37	6, 35 / 36, 37	1, 15 / 29, 4	28, 10 / 1, 39
8	Volume of a stationary object	-	35, 10 / 19, 14	19, 14	35, 8 / 2, 14	-	-	-	■	-	2, 18 / 37	24, 35	7, 2 / 35	34, 28 / 35, 40
9	Speed	2, 28 / 13, 38	-	13, 14 / 8	-	29, 30 / 34	-	7, 29 / 34	-	■	13, 28 / 15, 19	6, 18 / 38, 40	35, 15 / 18, 34	28, 33 / 1, 18
10	Force	8, 1 / 37, 18	18, 13 / 1, 28	17, 19 / 9, 36	28, 10	19, 10 / 15	1, 18 / 36, 37	15, 9 / 12, 37	2, 36 / 18, 37	13, 28 / 15, 12	■	18, 21 / 11	10, 35 / 40, 34	35, 10 / 21
11	Tension/Pressure	10, 36 / 37, 40	13, 29 / 10, 18	35, 10 / 36	35, 1 / 14, 16	10, 15 / 36, 28	10, 15 / 36, 37	6, 35 / 10	35, 24	6, 35 / 36	36, 35 / 21	■	35, 4 / 15, 10	35, 33 / 2, 40
12	Shape	8, 10 / 29, 40	15, 10 / 26, 3	29, 34 / 5, 4	13, 14 / 10, 7	5, 34 / 4, 10	-	14, 4 / 15, 22	7, 2 / 35	35, 15 / 34, 18	35, 10 / 37, 40	34, 15 / 10, 14	■	33, 1 / 18, 4
13	Stability of composition	21, 35 / 2, 39	26, 39 / 1, 40	13, 15 / 1, 28	37	2, 11 / 13	39	28, 10 / 19, 39	34, 28 / 35, 40	33, 15 / 28, 18	10, 35 / 21, 16	2, 35 / 40	22, 1 / 18, 4	■
14	Strength	1, 8, 40 / 15	40, 26 / 27, 1	1, 15 / 8, 35	15, 14 / 28, 26	3, 34 / 40, 29	9, 40 / 28	10, 15 / 14, 7	9, 14 / 17, 15	8, 13 / 26, 14	10, 18 / 3, 14	10, 3 / 18, 40	10, 30 / 35, 40	13, 17 / 35
15	Time of action of a moving object	19, 5 / 34, 31	-	2, 19 / 9	-	3, 17 / 19	-	10, 2 / 19, 30	-	3, 35 / 5	19, 2 / 16	19, 3 / 27	14, 26 / 28, 25	13, 3 / 35
16	Time of action of a stationary object	-	6, 27 / 19, 16	-	1, 40 / 35	-	-	-	35, 34 / 38	-	-	-	-	39, 3 / 35, 23
17	Temperature	36, 22 / 6, 38	22, 35 / 32	15, 19 / 9	15, 19 / 9	3, 35 / 39, 18	35, 38	34, 39 / 40, 18	35, 6 / 4	2, 28 / 36, 30	35, 10 / 3, 21	35, 39 / 19, 2	14, 22 / 19, 32	1, 35 / 32
18	Brightness	19, 1 / 32	2, 35 / 32	19, 32 / 16	-	19, 32 / 26	-	2, 13 / 10	-	10, 13 / 19	26, 19 / 6	-	32, 30	32, 3 / 27
19	Energy spent by a moving object	12, 18 / 28, 31	-	12, 28	-	15, 19 / 25	-	35, 13 / 18	-	8, 35	16, 26 / 21, 2	23, 14 / 25	12, 2 / 29	19, 13 / 17, 24
20	Energy spent by a stationary object	-	19, 9 / 6, 27	-	-	-	-	-	-	-	36, 37	-	-	27, 4 / 29, 18

Table 3-a: Altshuller's Table of Contradictions (Features to Improve 1-20 vs. Undesired Result 1-13)

Characteristics		Characteristic that is getting worse												
		14	15	16	17	18	19	20	21	22	23	24	25	26
1	Weight of a mobile object	28, 27 18, 40	5, 34 31, 35	-	6, 29 4, 38	19, 1 32	35, 12 34, 31	-	12, 36 18, 31	6, 2 34, 19	5, 35 3, 31	10, 24 35	10, 35 20, 28	3, 26 18, 31
2	Weight of a stationary object	28, 2 10, 27	-	2, 27 19, 6	28, 19 32, 22	19, 32 35		18, 19 28, 1	15, 19 18, 22	18, 19 28, 15	5, 8 13, 30	10, 15 35	10, 20 35, 26	19, 6 18, 26
3	Length of a mobile object	8, 35 29, 34	19	-	10, 15 19	32	8, 35 24	-	1, 35	7, 2 35, 39	4, 29 23, 10	1, 24	15, 2 29	29, 35
4	Length of a stationary object	15, 14 28, 26	-	1, 40 35	3, 35 38, 18	3, 25	-	-	12, 8	6, 28	10, 28 24, 35	24, 26	30, 29 14	-
5	Area of a mobile object	3, 15 40, 14	6, 3	-	2, 15 16	15, 32 19, 13	19, 32	-	19, 10 32, 18	15, 17 30, 26	10, 35 2, 39	30, 26	26, 4	29, 30 6, 13
6	Area of a stationary object	40	-	2, 10 19, 30	35, 39 38	-	-	-	17, 32	17, 7 30	10, 14 18, 39	30, 16	10, 35 4, 18	2, 18 40, 4
7	Volume of a mobile object	9, 14 15, 7	6, 35 4	-	34, 39 10, 18	2, 13 10	35	-	35, 6 13, 18	7, 15 13, 16	36, 39 34, 10	2, 22	2, 6 34, 10	29, 30 7
8	Volume of a stationary object	9, 14 17, 15	-	35, 34 38	35, 6 4			-	30, 6	-	10, 39 35, 34	-	35, 16 32, 18	35, 3
9	Speed	8, 3 26, 14	3, 19 35, 5	-	28, 30 36, 2	10, 13 19	8, 15 35, 38	-	19, 35 38, 2	14, 20 19, 35	10, 13 28, 38	13, 26	-	10, 19 29, 38
10	Force	35, 10 14, 27	19, 2	-	35, 10 21	-	19, 17 10	1, 16 36, 37	19, 35 18, 37	14, 15	8, 35 40, 5	-	10, 37 36	14, 29 18, 36
11	Tension/Pressure	9, 18 3, 40	19, 3 27	-	35, 39 19, 2	-	14, 24 10, 37	-	10, 35 14	2, 36 25	10, 36 3, 37	-	37, 36 4	10, 14 36
12	Shape	30, 14 10, 40	14, 26 9, 25	-	22, 14 19, 32	13, 15 32	2, 6 34, 14	-	4, 6 2	14	35, 29 3, 5	-	14, 10 34, 17	36, 22
13	Stability of composition	17, 9 15	13, 27 10, 35	39, 3 35, 23	35, 1 32	32, 3 27, 15	13, 19	27, 4 29, 18	32, 35 27, 31	14, 2 39, 6	2, 14 30, 40	-	35, 27	15, 32 35
14	Strength	■	27, 3 26	-	30, 10 40	35, 19	19, 35 10	35	10, 26 35, 28	35	35, 28 31, 40	-	29, 3 28, 10	29, 10 27
15	Time of action of a moving object	27, 3 10	■	-	19, 35 39	2, 19 4, 35	28, 6 35, 18	-	19, 10 35, 38	-	28, 27 3, 18	10	20, 10 28, 18	3, 35 10, 40
16	Time of action of a stationary object	-	-	■	19, 18 36, 40	-	-	-	16	-	27, 16 18, 38	10	28, 20 10, 16	3, 35 31
17	Temperature	10, 30 22, 40	19, 13 39	19, 18 36, 40	■	32, 30 21, 16	19, 15 3, 17	-	2, 14 17, 25	21, 17 35, 38	21, 36 29, 31	-	35, 28 21, 18	3, 17 30, 39
18	Brightness	35, 19	2, 19 6	-	32, 35 19	■	32, 1 19	32, 35 1, 15	32	13, 16 1, 6	13, 1	1, 6	19, 1 26, 17	1, 19
19	Energy spent by a moving object	5, 19 9, 35	28, 35 6, 18	-	19, 24 3, 14	2, 15 19	■	-	6, 19 37, 18	12, 22 15, 24	35, 24 18, 5	-	35, 38 19, 18	34, 23 16, 18
20	Energy spent by a stationary object	35	-	-	-	19, 2 35, 32	-	■	-	-	28, 27 18, 31	-	-	3, 35 31

(left margin label: Characteristic to be improved)

Table 3-b: Altshuller's Table of Contradictions (cont.) (Features to Improve 1-20 vs. Undesired Result 14-26)

Characteristics		Characteristic that is getting worse												
		27	28	29	30	31	32	33	34	35	36	37	38	39
1	Weight of a mobile object	3, 11 1, 27	28, 27 35, 26	28, 35 26, 18	22, 21 18, 27	22, 35 31, 39	27, 28 1, 36	35, 3 2, 24	2, 27 28, 11	29, 5 15, 8	26, 30 36, 34	28, 29 26, 32	26, 35 18, 19	35, 3 24, 37
2	Weight of a stationary object	10, 28 8, 3	18, 26 28	10, 1 35, 17	2, 19 22, 37	35, 22 1, 39	28, 1 9	6, 13 1, 32	2, 27 28, 11	19, 15 29	1, 10 26, 39	25, 28 17, 15	2, 26 35	1, 28 15, 35
3	Length of a mobile object	10, 14 29, 40	28, 32 4	10, 28 29, 37	1, 15 17, 24	17, 15	1, 29 17	15, 29 35, 4, 7	1, 28 10	14, 15 1, 16	1, 19 26, 24	35, 1 26, 24	17, 24 26, 16	14, 4 28, 29
4	Length of a stationary object	15, 29 28	32, 28 3	2, 32 10	1, 18	-	15, 17 27	2, 25	3	1, 35	1, 26	26	-	30, 14 7, 26
5	Area of a mobile object	29, 9	26, 28 32, 3	2, 32	22, 33 28, 1	17, 2 18, 39	13, 1 26, 24	15, 17 13, 16	15, 13 10, 1	15, 30	14, 1 13	2, 36 26, 18	14, 30 28, 23	10, 26 34, 2
6	Area of a stationary object	32, 35 40, 4	26, 28 32, 3	2, 29 18, 36	27, 2 39, 35	22, 1 40	40, 16	16, 4	16	15, 16	1, 18 36	2, 35 30, 18	23	10, 15 17, 7
7	Volume of a mobile object	14, 1 40, 11	26, 28	25, 28 2, 16	22, 21 27, 35	17, 2 40, 1	29, 1 40	15, 13 30, 12	10	15, 29	26, 1	29, 26 4	35, 34 16, 24	10, 6 2, 34
8	Volume of a stationary object	2, 35 16	-	35, 10 25	34, 39 19, 27	30, 18 35, 4	35	-	1	-	1, 31	2, 17 26	-	35, 37 10, 2
9	Speed	11, 35 27, 28	28, 32 1, 24	10, 28 32, 35	1, 28 35, 23	2, 24 35, 21	35, 13 8, 1	32, 28 13, 12	34, 2 28, 27	15, 10 26	10, 28 4, 34	3, 34 27, 16	10, 18	-
10	Force	3, 35 13, 21	35, 10 23, 24	28, 29 37, 36	1, 35 40, 18	13, 3 36, 24	15, 37 18, 1	1, 28 3, 25	15, 1 11	15, 17 18, 20	26, 35 10, 18	36, 37 10, 19	2, 35	3, 28 35, 37
11	Tension/Pressure	10, 13 19, 35	6, 28 25	3, 35	22, 2 37	2, 33 27, 18	1, 35 16	11	2	35	19, 1 35	2, 36 37	35, 24	10, 14 35, 37
12	Shape	10, 40 16	28, 32 1	32, 30 40	22, 1 2, 35	35, 1	1, 32 17, 28	32, 15 26	2, 13 1	1, 15 29	16, 29 1, 28	15, 13 39	15, 1 32	17, 26 34, 10
13	Stability of composition	-	13	18	35, 24 30, 18	35, 40 27, 39	35, 19	32, 35 30	2, 35 10, 16	35, 30 34, 2	2, 35 22, 26	35, 22 39, 23	1, 8 35	23, 35 40, 3
14	Strength	11, 3	3, 27 16	3, 27	18, 35 37, 1	15, 35 22, 2	11, 3 10, 32	32, 40 28, 2	27, 11 3	15, 3 32	2, 13 25, 28	27, 3 15, 40	15	29, 35 10, 14
15	Time of action of a moving object	11, 2 13	3	3, 27 16, 40	22, 15 33, 28	21, 39 16, 22	27, 1 4	12, 27	29, 10 27	1, 35 13	10, 4 29, 15	19, 29 39, 35	6, 10	35, 17 14, 19
16	Time of action of a stationary object	34, 27 6, 40	10, 26 24	-	17, 1 40, 33	22	35, 10	1	1	2	-	25, 34 6, 35	1	20, 10 16, 38
17	Temperature	19, 35 3, 10	32, 19 24	24	22, 33 35, 2	22, 35 2, 24	26, 27	26, 27	4, 10 16	2, 18 27	2, 17 16	3, 27 35, 31	26, 2 19, 16	15, 28 35
18	Brightness	-	11, 15 32	3, 32	15, 19	35, 19 32, 39	19, 35 28, 26	28, 26 19	15, 17 13, 16	15, 1 19	6, 32 13	32, 15	2, 26 10	2, 25 16
19	Energy spent by a moving object	19, 21 11, 27	3, 1 32	-	1, 35 6, 27	2, 35 6	28, 26 30	19, 35	1, 15 17, 28	15, 17 13, 16	2, 29 27, 28	35, 36	32, 2	12, 28 35
20	Energy spent by a stationary object	10, 36 23	-	-	10, 2 22, 37	19, 22 18	1, 4	-	-	-	-	19, 35 16, 25	-	1, 6

(Left margin, rotated: "Characteristic to be improved")

Table 3-c: Altshuller's Table of Contradictions (cont.) (Features to Improve 1-20 vs. Undesired Result 27-39)

Characteristics		Characteristic that is getting worse												
		1	2	3	4	5	6	7	8	9	10	11	12	13
21	Power	8, 36 38, 31	19, 26 17, 27	1, 10 35, 37	-	19, 38	17, 32 13, 38	35, 6 38	30, 6 25	15, 35 2	26, 2 36, 35	22, 10 35	29, 14 2, 40	35, 32 15, 31
22	Waste of energy	15, 6 19, 28	19, 6 18, 9	7, 2 6, 13	6, 38 7	15, 26 17, 30	17, 7 30, 18	7, 18 23	7	16, 35 38	36, 38	-	-	14, 2 39, 6
23	Waste of substance	35, 6 23, 40	35, 6 22, 32	14, 29 10, 39	10, 28 24	35, 2 10, 31	10, 18 39, 31	1, 29 30, 36	3, 39 18, 31	10, 13 28, 38	14, 15 18, 40	3, 36 37, 10	29, 35 3, 5	2, 14 30, 40
24	Loss of information	10, 24 35	10, 35 5	1, 26	26	30, 26	30, 16	-	2, 22	26, 32	-	-	-	-
25	Waste of time	10, 20 37, 35	10, 20 26, 5	15, 2 29	30, 24 14, 5	26, 4 5, 16	10, 35 17, 4	2, 5 34, 10	35, 16 32, 18	-	10, 37 36, 5	37, 36 4	4, 10 34, 17	35, 3 22, 5
26	Amount of substance	35, 6 18, 31	27, 26 18, 35	29, 14 35, 18	-	15, 14 29	2, 18 40, 4	15, 20 29	-	35, 29 34, 28	35, 14 3	10, 36 14, 3	35, 14	15, 2 17, 40
27	Reliability	3, 8 10, 40	3, 10 8, 28	15, 9 14, 4	15, 29 28, 11	17, 10 14, 16	32, 35 40, 4	3, 10 14, 24	2, 35 24	21, 35 11, 28	8, 28 10, 3	10, 24 35, 19	35, 1 16, 11	-
28	Accuracy of measurement	32, 35 26, 28	28, 35 25, 26	28, 26 5, 16	32, 28 3, 16	26, 28 32, 3	26, 28 32, 3	32, 13 6	-	28, 13 32, 24	32, 2	6, 28 32	6, 28 32	32, 35 13
29	Accuracy of manufacturing	28, 32 13, 18	28, 35 27, 9	10, 28 29, 37	2, 32 10	28, 33 29, 32	2, 29 18, 36	32, 28 2	25, 10 35	10, 28 32	28, 19 34, 36	3, 35	32, 30 40	30, 18
30	Harmful factors acting on object	22, 21 27, 39	2, 22 13, 24	17, 1 39, 4	1, 18	22, 1 33, 28	27, 2 39, 35	22, 23 37, 35	34, 39 19, 27	21, 22 35, 28	13, 35 39, 18	22, 2 37	22, 1 3, 35	35, 24 30, 18
31	Harmful side effects	19, 22 15, 39	35, 22 1, 39	17, 15 16, 22	-	17, 2 18, 39	22, 1 40	17, 2 40	30, 18 35, 4	35, 28 3, 23	35, 28 1, 40	2, 33 27, 18	35, 1	35, 40 27, 39
32	Manufacturability	28, 29 15, 16	1, 27 36, 13	1, 29 13, 17	15, 17 27	13, 1 26, 12	16, 40	13, 29 1, 40	35	35, 13 8, 1	35, 12	35, 19 1, 37	1, 28 13, 27	11, 13 1
33	Convenience of use	25, 2 13, 15	6, 13 1, 25	1, 17 13, 12	-	1, 17 13, 16	18, 16 15, 39	1, 16 35, 15	4, 18 39, 31	18, 13 34	28, 13 35	2, 32 12	15, 34 29, 28	32, 35 30
34	Repairability	2, 27 35, 11	2, 27 35, 11	1, 28 10, 25	3, 18 31	15, 13 32	16, 25	25, 2 35, 11	1	34, 9	1, 11 10	13	1, 13 2, 4	2, 35
35	Adaptability	1, 6 15, 8	19, 15 29, 16	35, 1 29, 2	1, 35 16	35, 30 29, 7	15, 16	15, 35 29	-	35, 10 14	15, 17 20	35, 16	15, 37 1, 8	35, 30 14
36	Complexity of device	26, 30 34, 36	2, 26 35, 39	1, 19 26, 24	26	14, 1 13, 16	6, 36	34, 26 6	1, 16	34, 10 28	26, 16	19, 1 35	29, 13 28, 15	2, 22 17, 19
37	Complexity of control	27, 26 28, 13	6, 13 28, 1	16, 17 26, 24	26	2, 13 18, 17	2, 39 30, 16	29, 1 4, 16	2, 18 26, 31	3, 4 16, 35	36, 28 40, 19	35, 36 37, 32	27, 13 1, 39	11, 22 39, 30
38	Level of automation	28, 26 18, 35	28, 26 35, 10	14, 13 17, 28	23	17, 14 13	-	35, 13 16	-	28, 10	2, 35	13, 35	15, 32 11, 13	18, 1
39	Productivity	35, 26 24, 37	28, 27 15, 3	18, 4 28, 38	30, 7 14, 26	10, 26 34, 31	10, 35 17, 7	2, 6 34, 10	35, 37 10, 2	-	28, 15 10, 36	10, 37 14	14, 10 34, 40	35, 3 22, 39

(Left axis label: Characteristic to be improved)

Table 3-d: Altshuller's Table of Contradictions (cont.) (Features to Improve 21-39 vs. Undesired Result 1-13)

Characteristics		Characteristic that is getting worse												
		14	15	16	17	18	19	20	21	22	23	24	25	26
21	Power	26, 10, 28	19, 35, 10, 38	16	2, 14, 17, 25	16, 6, 19	16, 6, 19, 37	-	■	10, 35, 38	28, 27, 18, 38	10, 19	35, 20, 10, 6	4, 34, 19
22	Waste of energy	26	-	-	19, 38, 7	1, 13, 32, 15	-	-	3, 38	■	35, 27, 2, 37	19, 10	10, 18, 32, 7	7, 18, 25
23	Waste of substance	35, 28, 31, 40	28, 27, 3, 18	27, 16, 18, 38	21, 36, 39, 31	1, 6, 13	35, 18, 24, 5	28, 27, 12, 31	28, 27, 18, 38	35, 27, 2, 31	■	-	15, 18, 35, 10	6, 3, 10, 24
24	Loss of information	-	10	10	-	19	-	-	10, 19	19, 10	-	■	24, 26, 28, 32	24, 28, 35
25	Waste of time	29, 3, 28, 18	20, 10, 28, 18	28, 20, 10, 16	35, 29, 21, 18	1, 19, 26, 17	35, 38, 19, 18	1	35, 20, 10, 6	10, 5, 18, 32	35, 18, 10, 39	24, 26, 28, 32	■	35, 38, 18, 16
26	Amount of substance	14, 35, 34, 10	3, 35, 40	3, 35, 31	3, 17, 39	-	34, 29, 16, 18	3, 35, 31	35	7, 18, 25	6, 3, 10, 24	24, 28, 35	35, 38, 18, 16	■
27	Reliability	11, 28	2, 35, 3, 25	34, 27, 6, 40	3, 35, 10	11, 32, 13	21, 11, 27, 19	36, 23	21, 11, 26, 31	10, 11, 35	10, 35, 29, 39	10, 28	10, 30, 4	21, 28, 40, 3
28	Accuracy of measurement	28, 6, 32	28, 6, 32	10, 26, 24	6, 19, 28, 24	6, 1, 32	3, 6, 32	-	3, 6, 32	26, 32, 27	10, 16, 31, 28	-	24, 34, 28, 32	2, 6, 32
29	Accuracy of manufacturing	3, 27	3, 27, 40	-	19, 26	3, 32	32, 2	-	32, 2	13, 32, 2	35, 31, 10, 24	-	32, 26, 28, 18	32, 30
30	Harmful factors acting on object	18, 35, 37, 1	22, 15, 33, 28	17, 1, 40, 33	22, 33, 35, 2	1, 19, 32, 13	1, 24, 6, 27	10, 2, 22, 37	19, 22, 31, 2	21, 22, 35, 2	33, 22, 19, 40	22, 10, 2	35, 18, 34	35, 33, 29, 31
31	Harmful side effects	15, 35, 22, 2	15, 22, 33, 31	21, 39, 16, 22	22, 35, 2, 24	19, 24, 39, 32	2, 35, 6	19, 22, 18	2, 35, 18	21, 35, 2, 22	10, 1, 34	10, 21, 29	1, 22	3, 24, 39, 1
32	Manufacturability	1, 3, 10, 32	27, 1, 4	35, 16	27, 26, 18	28, 24, 27, 1	28, 26, 27, 1	1, 4	27, 1, 12, 24	19, 35	15, 34, 33	32, 24, 18, 16	35, 28, 34, 4	35, 23, 1, 24
33	Convenience of use	32, 40, 3, 28	29, 3, 8, 25	1, 16, 25	26, 27, 13	13, 17, 1, 24	1, 13, 24	-	35, 34, 2, 10	2, 19, 13	28, 32, 2, 24	4, 10, 27, 22	4, 28, 10, 34	12, 35
34	Repairability	11, 1, 2, 9	11, 29, 28, 27	1	4, 10	15, 1, 13	15, 1, 28, 16	-	15, 10, 32, 2	15, 1, 32, 19	2, 35, 34, 27	-	32, 1, 10, 25	2, 28, 10, 25
35	Adaptability	35, 3, 32, 6	13, 1, 35	2, 16	27, 2, 3, 35	6, 22, 26, 1	19, 35, 29, 13	-	19, 1, 29	18, 15, 1	15, 10, 2, 13	-	35, 28	3, 35, 15
36	Complexity of device	2, 13, 28	10, 4, 28, 15	-	2, 17, 13	24, 17, 13	27, 2, 29, 28	-	20, 19, 30, 34	10, 35, 13, 2	35, 10, 28, 29	-	6, 29	13, 3, 27, 10
37	Complexity of control	27, 3, 15, 28	19, 29, 39, 25	25, 34, 6, 35	3, 27, 35, 16	2, 24, 26	35, 38	19, 35, 16	19, 1, 16, 10	35, 3, 15, 19	1, 18, 10, 24	35, 33, 27, 22	18, 28, 32, 9	3, 27, 29, 18
38	Level of automation	25, 13	6, 9	-	26, 2, 19	8, 32, 19	2, 32, 13	-	28, 2, 27	23, 28	35, 10, 18, 5	35, 33	24, 28, 35, 30	35, 13
39	Productivity	29, 28, 10, 18	35, 10, 2, 18	20, 10, 16, 38	35, 21, 28, 10	26, 17, 19, 1	35, 10, 38, 19	1	35, 20, 10	28, 10, 29, 35	28, 10, 35, 23	13, 15, 23	-	35, 38

Table 3-e: Altshuller's Table of Contradictions (cont.) (Features to Improve 21-39 vs. Undesired Result 14-26)

Characteristics		Characteristic that is getting worse												
		27	28	29	30	31	32	33	34	35	36	37	38	39
21	Power	19, 24 / 26, 31	32, 15 / 2	32, 2	19, 22 / 31, 2	2, 35 / 18	26, 10 / 34	26, 35 / 10	35, 2 / 10, 34	19, 17 / 34	20, 19 / 30, 34	19, 35 / 16	28, 2 / 17	28, 35 / 34
22	Waste of energy	11, 10 / 35	32	-	21, 22 / 35, 2	21, 35 / 2, 22	-	35, 32 / 1	2, 19	-	7, 33	35, 3 / 15, 23	2	28, 10 / 29, 35
23	Waste of substance	10, 29 / 39, 35	16, 34 / 31, 18	35, 10 / 24, 31	33, 22 / 30, 40	10, 1 / 34, 29	15, 34 / 33	32, 28 / 2, 24	2, 35 / 34, 27	15, 10 / 2	35, 10 / 28, 24	35, 18 / 10, 13	35, 10 / 18	28, 35 / 10, 23
24	Loss of information	10, 28 / 23	-	-	22, 10 / 1	10, 21 / 22	32	27, 22	-	-	-	35, 33	35	13, 23 / 15
25	Waste of time	10, 30 / 4	24, 34 / 28, 32	24, 26 / 28, 18	35, 18 / 34	35, 22 / 18, 39	35, 28 / 34, 4	4, 28 / 10, 34	32, 1 / 10	35, 28	6, 29	18, 28 / 32, 10	24, 28 / 35, 30	-
26	Amount of substance	18, 3 / 28, 40	13, 2 / 28	33, 30	35, 33 / 29, 31	3, 35 / 40, 39	29, 1 / 35, 27	35, 29 / 25, 10	2, 32 / 10, 25	15, 3 / 29	3, 13 / 27, 10	3, 27 / 29, 18	8, 35	13, 29 / 3, 27
27	Reliability	■	32, 3 / 11, 23	11, 32 / 1	27, 35 / 2, 40	35, 2 / 40, 26	-	27, 17 / 40	1, 11	13, 35 / 8, 24	13, 35 / 1	27, 40 / 28	11, 13 / 27	1, 35 / 29, 38
28	Accuracy of measurement	5, 11 / 1, 23	■	-	28, 24 / 22, 26	3, 33 / 39, 10	6, 35 / 25, 18	1, 13 / 17, 34	1, 32 / 13, 11	13, 35 / 2	27, 35 / 10, 34	26, 24 / 32, 28	28, 2 / 10, 34	10, 34 / 28, 32
29	Accuracy of manufacturing	11, 32 / 1	-	■	26, 28 / 10, 36	4, 17 / 34, 26	-	1, 32 / 35, 23	25, 10	-	26, 2 / 18	-	26, 28 / 18, 23	10, 18 / 32, 39
30	Harmful factors acting on object	27, 24 / 2, 40	28, 33 / 23, 26	26, 28 / 10, 18	■	-	24, 35 / 2	2, 25 / 28, 39	35, 10 / 2	35, 11 / 22, 31	22, 19 / 29, 40	22, 19 / 29, 40	33, 3 / 34	22, 35 / 13, 24
31	Harmful side effects	24, 2 / 40, 39	3, 33 / 26	4, 17 / 34, 26	-	■	-	-	-	-	19, 1 / 31	2, 21 / 27, 1	2	22, 35 / 18, 39
32	Manufacturability	-	1, 35 / 12, 18	-	24, 2	-	■	2, 5 / 13, 16	35, 1, 25 / 11, 9	2, 13 / 15	27, 26 / 1	6, 28 / 11, 1	8, 28 / 1	35, 1 / 10, 28
33	Convenience of use	17, 27 / 8, 40	25, 13 / 2, 34	1, 32 / 35, 23	2, 25 / 28, 39	-	2, 5 / 12	■	12, 26 / 1, 32	15, 34 / 1, 16	32, 26 / 12, 17	-	1, 34 / 12, 3	15, 1 / 28
34	Repairability	11, 10 / 1, 16	10, 2 / 13	25, 10	35, 10 / 2, 16	-	1, 35 / 11, 10	1, 12 / 26, 15	■	7, 1 / 4, 16	35, 1, 25 / 13, 11	-	34, 35 / 7, 13	1, 32 / 10
35	Adaptability	35, 13 / 8, 24	35, 5 / 1, 10	-	35, 11 / 32, 31	-	1, 13 / 31	15, 34 / 1, 16, 7	1, 16 / 7, 4	■	15, 29 / 37, 28	1	27, 34 / 35	1, 35 / 6, 37
36	Complexity of device	13, 35 / 1	2, 26 / 10, 34	26, 24 / 32	22, 19 / 29, 40	19, 1	27, 26 / 1, 13	27, 9 / 26, 24	1, 13	29, 15 / 28, 37	■	15, 10 / 37, 28	15, 1 / 24	12, 17 / 28
37	Complexity of control	27, 40 / 28, 8	26, 24 / 32, 28	-	22, 19 / 29, 28	2, 21	5, 28 / 11, 29	2, 5	12, 26	1, 15	15, 10 / 37, 28	■	34, 21	35, 18
38	Level of automation	11, 27 / 32	28, 26 / 10, 34	28, 26 / 18, 23	2, 33	2	1, 26 / 13	1, 12 / 34, 3	1, 35 / 13	27, 4 / 1, 35	15, 24 / 10	34, 27 / 25	■	5, 12 / 35, 26
39	Productivity	1, 35 / 10, 38	1, 10 / 34, 28	18, 10 / 32, 1	22, 35 / 13, 24	35, 22 / 18, 39	35, 28 / 2, 24	1, 28 / 7, 19	1, 32 / 10, 25	1, 35 / 28, 37	12, 17 / 28, 24	35, 18 / 27, 2	5, 12 / 35, 26	■

(Left vertical label: Characteristic to be improved)

Table 3-f: Altshuller's Table of Contradictions (cont.) (Features to Improve 21-39 vs. Undesired Result 27-39)

Substance-Field Analysis is one of TRIZ analytical tools. It can be used in the solution of problems related to technical or design activities through functional models building [1].

Substance-Field Analysis is a useful tool for identifying problems in a technical system and finding innovative solutions to these identified problems. Recognized as one of the most valuable contributions of TRIZ, Substance-Field Analysis is able to model a system in a simple graphical approach, to identify problems and also to offer standard solutions for system improvement [7].

The process of functional models construction comprehends the following stages [8]:

1. Survey of available information.

2. Construction of Substance-Field diagram.

3. Identification of problematic situation.

4. Choice of a generic solution (standard solution).

5. Development of a specific solution for the problem.

There are mainly five types of relationships among the substances: useful impact, harmful impact, excessive impact, insufficient impact and transformation [8].

Substance-Field Analysis has 76 standard solutions categorized into five classes [9]:

• Class 1: Construct or destroy a substance-field (13 standard solutions)

• Class 2: Develop a substance-field (23 standard solutions)

• Class 3: Transition from a base system to a super-system or to a subsystem (6 standard solutions)

• Class 4: Measure or detect anything within a technical system (17 standard solutions)

• Class 5: Introduce substances or fields into a technical system (17 standard solutions)

These 76 solutions can be condensed and generalized into seven standard solutions.

3. Practical cases of SF model application

An operation batch contains some pieces with characteristics out of specifications.

Figure 2 shows the problem (Problematic Situation 1 - Incomplete Model) [8].

Figure 2. Problematic Situation 1 - incomplete model

The Substance-Field Model is incomplete, a field is missing. The problem corresponds to Problematic Situation 1 and can be solved resorting to General Solution 1.

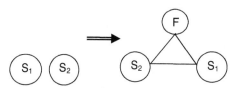

Figure 3. General Solution 1 for Problematic Situation 1

The possible specific solution is to inspect pieces before the operation, putting aside faulty components from acceptable ones. Then the model becomes complete.

A machine-tool fixture used for certain fabrication operation is damaging the lateral surfaces of the workpiece.

Figure 4 shows the problem (Problematic Situation 2 - Harmful Interactions between the Substances).

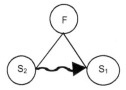

Figure 4. Problematic Situation 2 - harmful interactions between the substances

The Substance-Field Model is complete however the interaction between the substances is harmful. The problem corresponds to Problematic Situation 2 and can be solved resorting to General Solution 2.

Figure 5 shows the general solution.

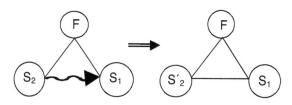

Figure 5. General Solution 2 for Problematic Situation 2

The possible specific solution is to use another machine-tool fixture system or to modify the actual fixture in order to eliminate or reduce damages at the lateral surfaces of the workpiece. Then the harmful interaction is reduced or eliminated.

General Solution 3: Modify S1 to be Insensitive or Less Sensitive to Harmful Impact

The problematic situation is the same (see Figure 4).

General Solution 3 is similar to General Solution 2, but instead of substance S2 modification, the substance S1 is modified. The characteristics (physical, chemical and/or other) of substance S1 are changed in order to become it less sensitive or insensitive to a harmful impact. The changes can be internal and/or external, can be temporary or permanent.

The physical and/or chemical characteristics of substance S1 may be altered internally or externally, so that it becomes less sensitive or insensitive to a harmful impact, as seen in Figure 4. The modification may be either temporary or permanent. Additives may be needed in the modification.

Figure 6 shows the general solution.

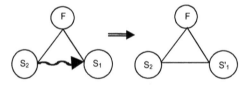

Figure 6. General Solution 3 for Problematic Situation 2

The possible specific solution is to create protection for the lateral surfaces of the workpiece. Then the harmful interaction is reduced or eliminated.

General Solution 4: Change Existing Field to Reduce or Eliminate Harmful Impact

The problematic situation is the same (see Figure 4).

General Solution 4 is similar to General Solutions 2 and 3, but instead of substances modification, the field F is modified.

Changing the existing field while keeping the same substances may be a choice to reduce or removing the harmful impact. The existing field can be increased, decreasing, or completely removed and replaced by another one.

Figure 7 shows the general solution.

The possible specific solution is to change the technological process and its operations keeping the same substances in order to reduce or eliminate the harmful interactions.

General Solution 5: Eliminate, Neutralize or Isolate Harmful Impact Using Another Counteractive Field F_x

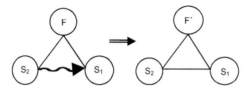

Figure 7. General Solution 4 for Problematic Situation 2

The problematic situation is the same (see Figure 4).

General Solution 5 presupposes introduction of a counteractive field F_X in order to remove, neutralize or isolate the harmful impact. The substances S_2 and S_1 and the field F will not change its characteristics in this solution.

Figure 8 shows the general solution.

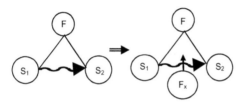

Figure 8. General Solution 5 for Problematic Situation 2

For example, a technological operation is creating significant superficial tensions in workpieces. The possible specific solution is to introduce a tempering operation (heat treatment) in order to reduce the superficial tensions.

General Solution 6: Introduce a Positive Field

The problematic situation is the same (see Figure 4).

General Solution 6 is very similar to General Solution 5.

Another field is added to work with the current field in order to increase the useful effect and reduce the negative effect of the existing system keeping all elements without change.

Figure 9 shows the general solution.

For example, Lean Philosophy is a systematized approach for continual improvement. The possible specific solution is to introduce another positive field, TRIZ techniques, so the useful effect of Lean is increases and negative effects in existing system are reduced.

General Solution 7: Expand Existing Substance-Field Model to a Chain

The problematic situation is the same (see Figure 4).

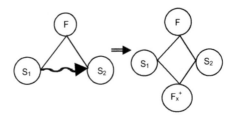

Figure 9. General Solution 6 for Problematic Situation 2

The existing Substance-Field Model can be expanded to a chain by introducing a new substance S3 to the system. Instead of directly acting upon S1, S2 will interact indirectly with S1 via another medium, substance S3.

Figure 10 shows the general solution.

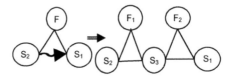

Figure 10. General Solution 7 for Problematic Situation 2

For example, it is difficult for a design team to obtain direct customer feedback about new product. The possible specific solution is to obtain customer feedback through the marketing and sales staff.

Beyond the Problematic Situation 1 (incomplete model) and the Problematic Situation 2 (harmful or undesirable interactions between the substances), also the Problematic Situation 3 (insufficient or inefficient impact) can occur.

Figure 11 shows the Problematic Situation 3.

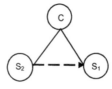

Figure 11. Problematic Situation 3 - insufficient or inefficient impact between the substances

The general solutions used for the Problematic Situation 2 can be used for the Problematic Situation 3. Figures 12-17 show the general solutions.

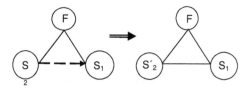

Figure 12. General Solution 2 for Problematic Situation 3

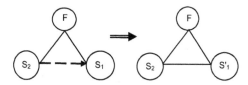

Figure 13. General Solution 3 for Problematic Situation 3

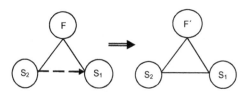

Figure 14. General Solution 4 for Problematic Situation 3

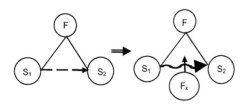

Figure 15. General Solution 5 for Problematic Situation 3

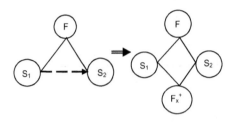

Figure 16. General Solution 6 for Problematic Situation 3

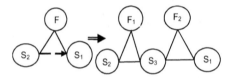

Figure 17. General Solution 7 for Problematic Situation 3

4. Ideality and application of ideality matrix to a camping stove case study

Consider the case of a camping stove design.

Customer requirements were collected, pooled and prepared by an affinity diagram, yielding the following list:

- Volume;
- Weight;
- Firing time;
- Noise level;
- Time required to boil water;
- Tank capacity;
- Burning time at maximum flame;
- Boiled water per unit of gas.

Table4 contains the Matrix of Ideality built for the camping stove.

Parameter	1.	2.	3.	4.	5.	6.	7.	8.
1. Volume		+			-	-	-	
2. Weight	+				-	-	-	
3. Firing time					+			
4. Noise level								
5. Time required to boil water	-	-	+				-	+
6. Tank capacity	-	-	+		+		+	+
7. Burning time at maximum flame	-	-			-	-		-
8. Boiled water per unit of gas	-	-			+	+	-	

- Harmful iteration

+ Useful iteration

Table 4. Ideality Matrix

The Ideality Matrix helps identify the interactions between the technical requirements and distinguish the positive and negative effects of iterations. For example, weight reduction can lead to reduction in volume, but may lead to reduction of the tank capacity.

Based on the Ideality Matrix, the level of ideality can be calculated as follows [10]:

$$\text{Ideality} = \text{Number of Useful Functions} / \text{Number of Harmful Functions} \tag{1}$$

In this case, the level of ideality is:

I= 11 / 30 ≈ 0,367

To increase the level of ideality it is necessary to move to the next phase, phase of solution of contradictions.

5. Application of ARIZ to a sterilizer case study

The AJC company runs its activity, since 1953, based on manufacturing medical and hospital material, being the main activity the conception, manufacture and assemblage of washer disinfectors of bed-pan and stainless steel utensils, vertical and horizontal steam sterilizers and steam generators.

The sterilization services implement in hospitals new philosophy which encompasses the traceability of equipment to use in the sterilization station, sterilized material, sterilization processes and handling operations with sterilized material and with material to be sterilized. The new sterilization philosophy leads to improvement of sterilizer capacity and sterilizer features. Actually hospitals need centralized management software for all sterilization equipment (including washing and disinfection machines, sterilizers, medical sealers), as well as materials to be sterilized (surgical, orthopaedic, textile and another utensils) and the sequence of operations (separation of material, washing and disinfecting, sorting and packing, sealing, sterilization, distribution and collection of material to be sterilized in the hospital).

The extant manufactured sterilizers are analyzed concerning its adaptation to the new tendencies of sterilization. ARIZ flowchart can be applied (Figure 18) [8].

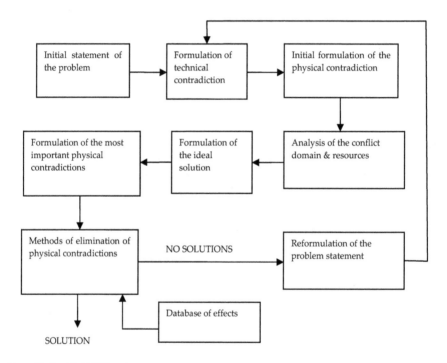

Figure 18. Simplified ARIZ flowchart

Several contradictions can be identified. It became clear that new sterilizers to be introduced on the market will have to undergo significant changes.

Beyond another features, the sterilizer must undergo changes at the level of pressure vessel where the sterilization of materials is performed.

The pressure vessel is constituted by the chamber, jacket, doors and other components welded to the pressure vessel.

The dimensions of the sterilization chamber can be modified. Former dimensions were: 70 centimeters of wide, 70 centimeters in height, 150 centimeters of depth with 735 liters of capacity. The new dimensions are: 70 centimeters in wide, 112 centimeters tall, 110 centimeters of depth with 862 liters of capacity.

The sterilizer door performance can be questioned too. The two doors working vertically can be replaced by two doors working horizontally.

The guide system for the door movement can be changed too. The doors can be sliding.

They detected a problem with difficult access to some mechanical and electrical components. The problem can be solved by modification of the component layout. The new sterilizers can have modular approach of the component layout; some mechanical and electrical components can be transferred to the lateral side of the sterilizer. Therefore, the maintenance operations will be easier.

They carried out the study of component layout (including electric framework, power framework, vacuum pump, water pump, condensate pan, valves, filters, etc.) to make assembly, maintenance intervention or replacement of the components easier.

As a result of all these changes, the structure that supports the sterilizer and its components can be reformed and redesigned too.

The sterilizer loading system can be undergone with important changes. The former loading system had an outside loading car that guided to the baskets load platform where the material to be sterilizes can be placed into the chamber. A similar outside loading car can be allowed to withdraw the load platform, baskets and utensils of the chamber in the clean area after sterilization (Figure 19a).

The loading system of the new sterilizer will be consists of a load car placed inside the chamber with the material to sterilize. There is a second outside loading car that transports the loading car from the sterilizer to the transportation board and vice-versa. The transportation board allows transport the sterilized material to outside and the material to be sterilized to the sterilization station (Figure 19b).

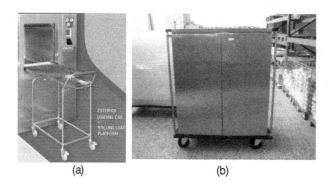

(a)   (b)

Figure 19. a) Former loading system; (b) New loading system

The Figure 20 shows two sterilizers:

• Sterilizer with sliding door vertically – before the application of TRIZ methodology (Fig. 20a).

• Sterilizer with sliding door horizontally – after the application of TRIZ methodology (Fig. 20b).

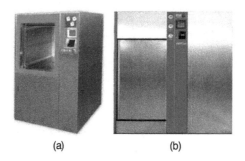

(a)   (b)

Figure 20. a) Former sterilizer AMARO 5000 with the door vertically; (b) New sterilizer AMARO 5000 with the sliding door horizontally

This project aims to achieve significant improvements in the quality of the sterilization process, simplifying the maintenance of the equipment, the traceability of the processes and equipment, the quality of work of the operators of sterilization station, making easier the sterilizer manufacture and assembly processes, the improvement and introduction of a new concept of charging system of the sterilizer. The developments described will allow company to submit an innovative concept that will be introduced in hospitals of medium and large dimensions.

6. Conclusions

The constant need for change, results in a current trend in industrial design activities. The industrial projects must achieve its objectives. Design teams need powerful and highly efficient analytical tools. One of the most important factors for the success of a project is the generation of ideas and innovation. The lack of creativity can lead to the failure of a project. Borderless communication, information and innovation are crucial for design competitiveness.

The TRIZ methodology, with its strong theme of innovation, can contribute to accelerating the resolution of problems in the industrial design activities [11]. The TRIZ analytical tools would be very useful for schematization of project tasks, structural analysis, identification and formalization of contradictions and problematical situations and its solving.

Acknowledgements

The present author would like to thank the Faculty of Science and Technology of The New University of Lisbon (UNL) and the Portuguese Foundation for Science and Technology (FCT)

through the Strategic Project no. PEst-OE/EME/UI0667/2011. Their support is helping to make possible our research work.

Author details

Helena V. G. Navas

UNIDEMI, Department of Mechanical and Industrial Engineering, Faculty of Science and Technology, New University of Lisbon, Portugal

References

[1] Altshuller, G. S. Creativity as an Exact Science: The Theory of the Solution of Inventive Problems, Gordon and Breach Publishers., (1995).

[2] Savransky, Seymon D., Engineering of Creativity: Introduction to TRIZ Methodology of Inventive Problem Solving, CRC Press, Boca Raton, Florida, (2000).

[3] Altshuller, G. S. Principles: TRIZ Keys to Technical Innovation, Technical Innovation Center, (2001).

[4] Fey, V. R, & Rivin, E. I. The Science of Innovation: A Managerial Overview of the TRIZ Methodology, TRIZ Group, Southfield, (1997).

[5] Radeka, K. TRIZ for Lean Innovation: Increase Your Ability to Leverage Innovation Across the Enterprise and Beyond, Whittier Consulting Group, Inc., (2007).

[6] Terninko, J, Zusman, A, & Zlotin, B. Systematic Innovation: An Introduction to TRIZ (Theory of Inventing Problem Solving), St. Lucie Press, (1998).

[7] Mao, X, Zhang, X, & Rizk, S. Generalized Solutions for Su-Field Analysis, The TRIZ Journal, August (2007).

[8] Altshuller, G. S. Tools of Classical TRIZ, Ideation International Inc., (1999).

[9] Terninko, J, Domb, E, & Miller, J. The Seventy-Six Standard Solutions, with Examples, The TRIZ Journal, February, March, May, June and July (2007).

[10] Domb, E. How to Help TRIZ Beginners Succeed, http://www.triz-journal.com/archives/1997/04/a/index.html,(1997).

[11] Coelho, D. A. Matching TRIZ Engineering Parameters to Human Factors Issues in Manufacturing, WSEAS Transactions on Business and Economics, November (2009). , 6(11), 547-556.

Chapter 4

Product Sound Design:
Intentional and Consequential Sounds

Lau Langeveld, René van Egmond,
Reinier Jansen and Elif Özcan

Additional information is available at the end of the chapter

1. Introduction

In our daily life we are immersed in sounds that are generated by products. If one were to ask someone to name sounds produced by products, often sounds are mentioned that alarm or inform us (e.g., microwave oven beeps, telephone rings etc.). These are the sounds of which we are consciously aware. However, many sounds subconsciously play an important role in our interaction with a product. One hears if the battery of a toothbrush runs out of power; one hears the power of a vacuum cleaner and one hears if the bag is full; etc. Although these are all functional aspects, sound also plays a role in our aesthetic, quality, and emotional experience of products. For example, one hears if the sound of a car door evokes a sense of quality. Car manufacturers have acoustical engineers to make sure that a slammed door will evoke this sense of quality. Sound quality and its relation to perception have been studied to some extent (e.g., Blauert & Jekosch, 1997; Bodden, 2000; Lyon, 2003). Often, these methodologies cover only one aspect of the design or evaluative process. Here we present a systematic approach to the inclusion of sound in the design process and its use as an essential aspect of controlling the quality of design and as a means of educating designers (and students) about the constituent parts of a product.

In this chapter, we will distinguish between sounds that are generated by the operating of the product itself and sounds that we intentionally add to a product. In the field of product sounds the first category has been named consequential sounds and the second category has been named intentional sounds (Van Egmond, 2007). This distinction is essential - both categories of sounds will require different design methods and the use of knowledge of different disciplines is needed. Intentional sounds are mostly composed which may be experienced as musical sounds. One could state that the use of intentional sounds as feedback of alarm sounds is in fact creating a small musical composition (i.e., musical motives). Therefore, these sounds can also be used to convey brand values of companies.

Consequential product sounds are experienced as "noisy". It is very difficult for users, for designers, and for acoustical engineers to verbally express how they experience a sound. Several problems exist. In general, users lack the vocabulary to express themselves to explain what is wrong or right with a sound. They normally will say the product makes a unpleasant sound or noise. Designers also lack the vocabulary to express design concepts that may also be used in the design of a sound. The acoustical engineers have a very technical vocabulary from the disciplines of physics and sometimes psychoacoustics, which does not communicate very well to designers and to users. In addition, to understand the aesthetic and emotional experience of product sounds knowledge from the field of psychology (auditory perception, cognition, and emotion theories) is needed. As stated before, product sounds are loud and noisy. This inherent property makes it difficult to describe the sound in a structural manner. The reason for this is, of course, that noise by itself is random and lacks structure. However, product sounds do not produce completely random noise due to the resonance and engine/boiler properties of products (of course, there are many sources that are responsible for the generation of sound in domestic appliances). It is the aim of this product sound course of Industrial design Engineering(IDE), Delft University of Technology (DUT) to try to relate descriptive aspects from the physical, perceptual, and experiential domain to each other in order to improve the sound of domestic appliances.

1.1. The perception of sound

The top-down processing (involving knowledge stored in memory or mental representations) will result in the attribution of meaning (e.g., recognition, identification), relating sound to certain events, evoking (cognitive) emotions. It is important to note that the sensorial experience of a sound can be — often — directly related to the spectral and temporal features of a sound, whereas this is more difficult for top-down aspects (except for very well-structured sounds like speech and music). As described above, one of the aspects that is well known is the irritation that sounds evoke. The irritation can often be contributed to the sensorial processing of the sounds. It can be argued that top-down aspects, like the attribution of meaning, can positively influence the experience whether a sound is irritating or not.

In courses, students hear the sound of an epilator. This sound evokes a rattling and rough experience. If the students are asked to tell the source of the sound most students say this sound stems from a hedge-trimmer or some other power tool. If they are told that it is an epilator and they listen to the sound for a second time, the look on their face is completely different and reveals a sense of unpleasantness. Thus, the experience of a sound changes if the meaning is known. One of the perceptual aspects that cause this is the rattling of the product caused by the construction, the gears, and the engine. This aspect can be captured by the measure of roughness. This attribute can be related to the structural properties of the sound in spectral and temporal domains and is one of the determinants in the perception of sensory pleasantness.

2. Products

A product is the result of a design process that starts with a design problem, involves ideation phases, and ultimately leads to a market introduction. In the context of product sound design, mainly domestic appliances are considered. The appliances have moving parts that can move linear or radial and are joined together in such a way to fulfills its functional aspect. and in particular the sound of the product. The product sound is influenced by many physical parameters such as: material, size, form, stiffness, load, energy etc.

2.1. Technology

Energy facilities are dependent on the place of use. For instance a product with a combustion engine is not used in houses or factory halls, because the pollution of the environment and sound intensity. Electricity is the most convenient energy type which is available in the form of batteries and power outlet. All other types of energy such as, hydro-electric power, fuel cells, human power, solar energy and atomic energy are not considered because the main power source is electricity. Electricity is easy to convert into another type of energy such as: thermo energy, mechanical energy, chemical energy, etc., but every conversion means an energy loss.

Product sounds manifest themselves in mainly three sources airborne sound, liquid sound and structure-borne sound. In a product we are dealing mainly with structure-borne sound sources that find their way to the outside environment by radiation. Transfer paths take care of the propagation of the sound from the source to the environment of the product. Structure-borne sound demonstrates itself in solids, in constructions that are built up from plates, beams, shells and shafts. The material properties determine the propagation speed, which is constant for certain waves and forms. The propagation speed depends on elasticity, specific gravity and contraction, which is different for solid materials. However, steel and aluminium have the same propagation speed because the division of the elasticity by the density is the same (E/ϱ).

Medium	Prolongation speed in m/s
Air	340
Water	1500
Steel, Aluminium	5200
Iron	5200
Brass	3700
Glass (window)	6800
Wood (parallel) $\rho=0.5$ kg/dm^3	4000
Lucite (Plexiglas)	2650
Polystyrene	2300
Rubber (soft)	50

Table 1. Propagation speed for a number of materials, liquid, and air (Verheij, 1992).

The size of a product is determined by the required function and power needed to fulfil the function. For instance, a toaster: the electricity is converted into heating power for toasting the bread. The size of a toaster depends on the efficiency of the reflection and isolation of the power. The heating element is a sound source. An additional sound source is the relief mechanism.

The bending stiffness is the elasticity (E) multiplied by the moment of inertia (I) which is dependent on the cross-section type and the dimensions. For instance an electric milk shaker has a bar with a certain mass on the end. If the bar turns around then it is bending under the gyroscopic force. Better bending stiffness could be achieved if a hollow profile is used. This is because the mass is further away from the centre of gravity. The bending energy is dependent on the stiffness but has a lower bending stiffness means a higher bending energy. This bending energy is transformed into sound.

A load is needed to fulfil the required mechanical function for a right performance of the appliance at a certain speed. The power required for the function fulfilling is the load a torque (T), that is necessary for processing times the speed (ω). The needed power out, Pout= ω x T. The power input multiplied by the efficiency of energy transforming η_{et} and mechanical transmission η_{mt} is, Pout = Pin x η_{et} x $\eta_{mt.}$. The choice of type of power and mechanical drive is really important for the overall efficiency η_{eff}. If the efficiency of the permanent magnetic motor is 50% and the mechanical drive 80% for each transmission in a three-step drive, then the efficiency is only 25.6%. The conclusion could be that the best drive is the one without the mechanical transmission, so the energy transforming is only responsible for the efficiency. The biggest advantage of a direct drive is less parts, which reduce enormously an amount of sound sources. The efficiency of a product depends on the energy losses which are transformed through friction into heat, and the movements of masses in sound or noise.

Moving product parts are necessary to fulfil the function of domestic appliances which have six degrees of freedom in a three dimensional space, three degrees are transversal and the other three are rotational. The complete drive of a domestic appliance consists of an electric motor with a mechanical transmission that will be built up with machined parts to adjust the revolutions per minute that is needed with a certain torque. The best drive is without any moving parts, the direct drive. For example, a gear shaft has only one degree of freedom, namely rotation around its own axis. All the other degrees of freedom will be restricted to zero by the construction. For this purpose, fixed, detachable, and combination joints are available. A certain clearance is needed to realize their relative movements. A minimum and a maximum clearance can be determined depending on the tolerances of two parts, as well as small expansions due to temperature increases. Tolerances are the result of the chosen manufacturing process, which is determined by material, shape, size and volume of production. The clearance in the joints of the parts has a certain freedom of movement of the masses. This costs an amount of moving energy which will be transformed in a product sound.

2.2. Product domains

The product has relationships with three domains: Design, Embodiment and Production (see figure 1). The domains have a relationship which three conditions: environment, designer, and manufacture. The relationship between the domains are the activities: Creating, Designing and Making.

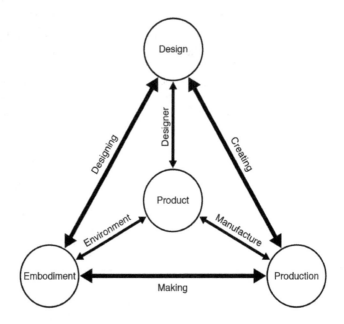

Figure 1. Figure 1: Product relationship with the domains production, embodiment, and design.

Each domain is associated with different levels see table 2. The domain design associates the user, observer and owner level. The domain product points out the physical, sensory and social experiences (after derivation of design from the consumer's point of view (Heufler, 2004). The two other domains have also pointed out the levels, the levels are for embodiment: practical, aesthetically, and status symbol. The levels are for production: parts making and assembly, availability, and benefit. The eye catcher is 'benefit' at the domain production, because without any profit, no other activity shall be undertaken which results in the production of products. Investments are made in production and design. Finally the activities should be profitable in a given period.

Design	Product	Embodiment	Production
User level	Physical experience	Practical	Parts making & Assembly
Observer level	Sensory experience	Aesthetically	Availability
Owner level	Social experience	Status symbol	Benefit

Table 2. Domains with their levels.

The relationship between design and production is the activity creating, or availability of production facilities. The availability is necessary to create the production under the constraints of the part design, manufacturing process and material. The designer must have good knowledge of: manufacturing and assembly, material, and aesthetics to create a successful product sound design.

The relationship between design and embodiment is the activity of designing, which is mostly carried out by a designer. Note that, designing is not engineering but creative problem solving, which always results in an embodiment. Engineering is a structured way of solving problems which lead to technical solutions that result in, objects, or systems.

The relationship between embodiment and production is the activity of making, the realization of a product with machinery or manually with a set of tools. Making gives satisfaction to a designer that a product can be realized. This experience is increasing the personality and identity of designer. The conditions form the relationship of the product with the domains of design, production and embodiment. Manufacturers make it possible to realize products by means of production This condition is an important connection in the product realization process. The designer is the condition to realize the product from a design. However, there are differences in the quality of the product design. This is caused by the personality and identity of a designer. The environment is the condition that an embodiment can be manifested as a product. The product should be manufactured from raw material to parts which will be assembled as a whole. This may be a component, a sub-assembly or a product. Two kinds of tolerances occurred, which are known as dimensional tolerances and geometric tolerances, in manufacturing of parts and in the assembly of parts into a whole. Every manufacturing process has its own tolerances that depends on material, type of process, stiffness, geometry etc. For steel and aluminium the ratio E/ϱ is almost equal, so also sound prolongation speed. Elastic modulus is always influencing the stiffness coefficient EI with I as moment of inertia, which results in dimensional tolerances and geometric tolerances. However, the manufacturing process speed and force have influence on the size of tolerance, but not on geometric (form). The power of the manufacturing process is transformed is force and speed, which the temperature of the work will rise higher. The height of temperature is dependent on the power needed for the process and processed material. It results in temperature elongation, which influenced the tolerances after cooling of the work to the temperature of the environment.

The designers make the manufacturing process choices which depends on material, shape, size and volume of production, which results in certain dimensional and geometric accuracy. After assembling two parts, the clearance will manifest as a result of the separate accuracy of these parts. For instance for plastics it is harder to reach the accuracy, because the temperature elongation is much higher than that of steel. For instance a folding plastic garden chair should be able to fold up, which is made possible by hinge points which are required if the parts have to move freely. The chair does not have a power source to conduct, but has a force to hold. Here, large tolerances are acceptable while the comfort is not being affected. With plastic, these tolerances are achievable despite the poor accuracy of the manufacturing processes. Shape of a part is to be achieved by cutting, extrusion, forging, moulding, casting, stamping, forming etc. However, not every material is applicable for every manufacturing process. But the size always has limitations resulting from the starting material, for example wood is limited by the age of the tree. The volume of production can range from single pieces to mass; this requires constantly changing of the manufacturing processes and thus different clearance requirements are possible. At mass production, the tolerances are under control; otherwise the failure rate is too high. Zero defects is possible with mass production. However, before this is achieved, the entire production system must be calibrated.

The manufacture makes the parts between the upper and lower limits of the tolerance. The clearance between two assembled parts will be between maximum and minimum size of the individual components. A minimum clearance is preferred because the excitation has than the smallest movement, and the smallest influence on the components, resulting in a lower sound pressure. Of course it is unique to reach this situation by means of a manufacturing system. Most clearances are reached between averages of the tolerances. Every domestic appliances produce sound. The production of these sounds is a consequence of their operating and construction. Therefore, these sounds are called consequential sounds. These sounds should be analysed in the physical, perceptual, and emotional domain to relate subjective findings to the engineered parts of the product.

If a domestic appliance is switched on then the power will conduct through the construction of parts to fulfil the working principle. Efficiency of the function is never hundred percent the losses are raised by friction of moving parts and vibration of parts, by mechanical excitation of the construction.

2.3. Consequential product sound model

The consequential product sound model are shown in figure 2, with four main aspects: sources, transmission of sound in the product, radiation, and transmission to receiver.

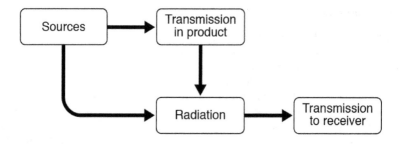

Figure 2. Model for product sound.

The sources of sound are defined as: airborne sound, liquid borne sound and structure borne sound. Gaver (1993) mentioned the events as sound sources, the interaction of material at a location in an environment with a certain impact caused by the power. The power sources could be from outside the product as electricity, water, gas and air. For example electricity is mostly used in domestic appliances or consumer goods such as: coffee maker, dish washer, extractor fan, convection oven, electric drill, shaver, grinder, hairdryer etc. Examples for water could be the tap in the kitchen, water sprinkler for the garden, sprinkler installation as fire protection etc. Gas and air also have good examples in the home such as: stove in the kitchen, airbrushes for decoration etc.

The energy can also be stored in a battery, gas bottle, container or human. This energy can be delivered to the product at the desired moment for a limited time. For example, the water tank (container) of a toilet is used for flushing the toilet bowl after use of the toilet and it is then filled again. The water contains the amount of potential energy that is needed to flush the toilet.

In table 3 (next page) the sound sources with primary excited medium are defined. Type of excitation should be always an activity such as: mechanical, aero dynamical, hydro dynamical etc. The examples are experienced in daily life in a household and a manufacturing plant. Özcan has six product sound categories defined; these are not based on sources but on the experience of the sources. There is always energy stored in the sources or fed from outside the product by the power outlet.

Radiation is the excitation of airborne sound by surfaces and other parts of a product. In water such as: mobility for boats, water bikes, wind surfboard, etc. the radiation of sound is also important. Transmission of sound takes place by means of the transfer of the primary excited medium such as construction, air and liquid. In a product multiple propagation paths may occur depending on the product layout. Construction sound transmission are carried out by the components of the product, but air and liquid sound transmission is carried out by air and liquid -filled cavities or by the mediums air and liquid.

The receiver always experiences the product sound in an environment, but the sound propagates from the product by air. However, the sources are experienced after the transmission in the product and the radiation to the environment. Cooking on a gas stove is nice example because you experience the amount of gas flows that simultaneous burns. Besides you also experience the gas flow as high or low at a certain distance. The gas supply with small pressure and combustion have an interaction with the environment.

Two approaches are possible to create the desired product sound. The first approach may reduce noise, e.g., projector, air conditioning, air hammer etc. The second approach is a product sound design (powerful experience, intensive experience) to be designed; e.g., electric shaver, toothbrush, electric power tool.

Before the desired product sound can be designed, the product must be measured against the sound of an existing product. From a product design, a prototype can be built which can also be measured. Measuring the individual contributions of the parts and components are notified through disassembling (deconstruction) a product, removing part after part.

Sources	Primary excited medium	Type of excitation	Examples
Air borne sound	Air or another gas	Mechanical	Compressor Refrigerator
		Aero dynamical	Fan Turbulent flow
		Combustion	Exhaust gasses in the exhaust pipe Autogenous welding Gas burner
Liquid borne sound	Liquid	Mechanical	Plunger pump Gear pump
		Hydro dynamical	Turbulence in flow Cavitation
Structure borne sound	Construction	Mechanical	Inertia: unbalanced
		Aero dynamical	Collide: hammering, rolling, stamping, sawing Turbulence gas flow Air spray
		Hydro dynamical	Releasing of whirls Air jet on surface Water heating
		Electro mechanical	Pole attraction Magnetostriction in transformer

Table 3. Defining the sound sources.

3. Intentional product sounds

Intentional sounds are 'intentionally' implemented and are typically produced by means of a loudspeaker or piezo element. They are mostly digital and somewhat musical sounds often used in user interfaces. Intentional sounds can be found in, e.g., domestic appliances (e.g., alarm clocks, mobile phone button beeps, microwave oven finish bells, operating system welcome tunes), automotive (e.g., low fuel warning, unfastened seatbelt alert), public transport (e.g., beeps at check-in points), and healthcare (e.g., heart-rate monitoring). These synthesized or recorded sounds are typically created using music software. The function of intentional sounds is often to alarm or to provide feedback to users.

This section first provides an elaboration on different functions and types of intentional sounds. Then, an overview will be given on commonly used techniques for implementation. A suggested design process for these sounds can be found later in this chapter.

3.1. Functions of intentional sound

Added sounds are regularly used to communicate abstract meanings or to provide information about the result of a process or activity (feedback). For example, when pressing membrane buttons on a microwave oven, the buttons themselves do not make sound. However, a 'beep' sound produced by a built-in piezo element will confirm the user's choice, after which the microwave's platform starts to rotate and produce its typical cyclic sound. This illustrates how as augmentation, intentional sounds are not inherently coupled to either a user's action or a product's functionality (see: consequential sounds). Yet, listeners learn to attribute meaning to added sounds, as they are generally designed to convey certain messages. For example, Edworthy *et al.* (1995) investigated the potential effects of changes in acoustic parameters (e.g., pitch, rhythm) on associated meanings (e.g., controlled, dangerous, steady). This attribution process is highly context-dependent. Consider how the perceived urgency of identical warning sounds may be different depending on whether it indicates a low battery warning of a mobile phone, or a problem with a heart rate monitoring system. See Hoggan *et al.* (2009) for an example on contextual differences in mapping audio parameters to informing signals by user interfaces (i.e., confirmations, errors, progress updates, warnings). Furthermore, product sounds are always part of a larger auditory environment. For example, an intensive care unit consists of a wide range of monitoring equipment. Lacking a standard for their alarm sounds, nurses potentially mistake a 'code red' alarm of one machine for a 'mild' alarm of another machine (Freudenthal *et al.*, 2005, Sanderson *et al.* 2009). Therefore, it is essential to design intentional sounds based on the interactions users (should) have with the product in a given context, and based on how people perceive these sounds.

One can differentiate between discrete and continuous feedback. The button tones of a microwave oven serve as confirmation of a completed action. They give discrete feedback, as they only sound once after a key has been pressed. This is different from continuous monitoring of a process, such as the series of beeps emitted by parking assistants in modern cars. Here, the time between consecutive beeps is inversely related to the distance to the car behind. Therefore, this is also an example of dynamic feedback. On the other hand, the microwave button tones always sound the same, regardless of how the user pushes them. Thus, this type of feedback can be called static. The decisions between discrete vs. continuous and dynamic vs. static feedback have consequences on the implementation of the corresponding sounds, as will be shown later.

3.2. Classes of intentional sound

One can discern between four main classes of intentional sounds: earcons, auditory icons, sonification, and continuous sonic interaction. The examples given so far mainly consisted of beep-like sounds. They are part of a larger class of discrete musical sounds which are called *earcons*. As discussed before, the abstract mapping of earcons must be learned, as there is no semantic link between the sounds and the data they represent. Differentiation is commonly found in terms of pitch, rhythm, timbre, spatial location, duration, and tempo (Hoggan *et al.* 2009). A second class of intentional sounds are *auditory icons*. Contrary to earcons, these are natural, everyday sounds, which are described in terms of their sources (e.g., the air flow sound of a fan to represent the state of a steam vent). Due to their semantic link to the things they represent, auditory icons are supposedly easier to learn and remember than earcons (Hearst *et al.* 1997). A third class of intentional sound is *sonification*, which concerns continuous data display. An ongoing awareness of a total system can be created, by including both alarming sounds and reassuring sounds for 'normal' states. Barrass argues that sonification can be used for monitoring an entire system, whereas earcons and auditory icons are better suited for diagnosis of subsystems (Hearst *et al.* 1997). Finally, a fourth class of intentional sounds has emerged. Rather than focusing on system states, *continuous sonic interaction* aims at sonifying expressiveness in human-product-interaction. A study by Rocchesso *et al.* (2009) illustrates how dynamic, continuous sound can influence the way we interact with a range of experimental kitchen appliances.

3.3. Implementation of intentional sounds

Intentional product sounds are typically generated with music software. The type of implementation depends on the classes of intentional sounds. Two main approaches can be discerned: recording and parametric synthesis. In the recording approach, (parts of a) product or environment are recorded, which can be done outdoors with a field recorder, or in an acoustically-treated recording room. The absence of room reverb in the latter condition facilitates editing at a later time. Recordings can be manipulated (e.g., equalization, compression), sliced, and layered to create a more complex sound. The main advantage of using recordings is the ease with which a realistic sound can be obtained. This approach lends itself well to creating auditory icons. However, recordings are not as flexibly manipulated as sounds created with parametric synthesis.

Parametric sound synthesis concerns the creation of sound starting from nothing. This implies that every sound feature deemed important should be included in a model. With such a model, the sound can then be manipulated according to its corresponding parameters. Typical techniques include additive, subtractive, wavetable, amplitude modulation, frequency modulation, and granular synthesis (examples on these techniques can be found in Farnell, 2010). Here, the use of elementary waveforms (i.e., sine, saw tooth, triangle) and/or noise is the common starting point. Parameters usually relate to an acoustical description of the sound (e.g., saw tooth wave and filter cutoff frequencies). Another technique that has gained increased attention over the years is physical modelling. This technique commonly employs mass-spring, dampener, and resonator models that mimic the working principles and construction of, e.g., musical instruments. Consequentially, parameters relate to 'natural' features, such as plucking force, string length, and material thickness. Rocchesso *et al.* (2009) argue that for continuous sonic interaction "the main sound design problem is not that of finding which sound is appropriate for a given gesture. Instead, the main problem is that of finding a sensible fitting of the interaction primitives with the dynamical properties of some sound models, in terms of specific perceptual effects." Parametric synthesis offers great flexibility, but at the cost of an increased effort to generate realistic, appropriate sounds.

The product sound designer should decide whether the sound will be presented static or dynamic. In the case of static sounds, one may choose to save them as samples to a dedicated piece of memory. The samples can then be played back on-demand. This is often the case with auditory icons and earcons. However, for sonification and continuous sonic interaction, both dynamic by definition, the synthesis model itself will have to be implemented in the chipset of the product. The sound will then be generated and manipulated in real-time, depending on the input of sensors. Note: the implementation of a synthesis model is not always feasible for complex sounds that require CPU processor-intensive models.

Finally, a sound that has been created digitally requires at least a digital-to-analogue convertor, and a loudspeaker or piezo element to be heard. For optimal acoustic efficiency, the resonance frequency of the cavity in which the loudspeaker or piezo element is mounted may require tuning to the frequency content of the envisioned sound.

4. Product sound design process

Aforementioned intentional and consequential sounds can be designed in order to facilitate a certain product experience. The main aim of the sound design process is to facilitate an auditory experience by using product sounds that are complimentary or supportive to the main product experience. For example, the warning signal of a microwave oven could be designed to be 'inviting' or a shaver could be designed to sound 'sporty'. In both examples, the desired auditory experience can only be achieved by forcing changes into the constructive elements of the main product, as sound is a natural consequence of objects/materials in action.

The design of the consequential and intentional sounds undergoes an iterative process (similar to the method suggested by Roozenburg and Eekels, 2003) that runs parallel to the main design process so that communication between different design teams is kept at its highest level of knowledge-exchange. Thus, a product sound design process incorporates four stages (see Figure 3):

1. *sound analysis* within product usage context;

2. *conceptualization* of ideas with sounding sketches;

3. *embodiment* of the concept with working and sounding prototypes;

4. *detailing* of the product for manufacturing with sounds fine-tuned to their purpose.

In light of the four-stage sound design process, it is often the case that sound design process starts with the main design brief, in which special attention may have been paid specifically to sound. However, usually the main design concept suggested in the brief can be taken as the basis for sound design.

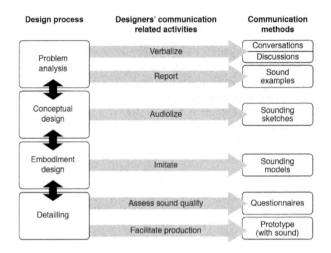

Figure 3. Methods for product sound design -related activities (adapted from: Özcan & van Egmond (2008))

4.1. Stage 1: Sound analysis

The sound analyses stage starts by first determining when and how the product emits sound and how the sound is incorporated into the human-product interaction. Therefore, observational research with high-definition audio-visual recordings is necessary to place the sound in context with the user in an environment natural to human-product interactions. In such observations, the following issues should be considered or paid attention to:

- acoustic effect of environment on the sound,

- other environmental/product-related sounds that could mask the sound in question,

- interaction of the product with the user and environment,

- facial expressions of users for detecting unpleasant or unwanted sounds,

- stages of product use and occurrence of sound in any given stage,

- duration of the product use and exposure to sound,

- impact of sound on product usability.

After tackling these issues and making a map of auditory experience within context, dry recordings of the product sound in a studio environment can be taken. Both dry and environmental sound recordings can be further analysed in terms of acoustic content of the sounds (e.g., Spectrograms, Bark scales) and their basic relevance to psychoacoustics. Subsequently, a comparison can be graphically made between a product sound occurring in a natural environment and the actual sound of the product without any environmental effects.

The acoustical analysis of sounds is also used to pinpoint acoustic regions that can cause sensory discomfort and locate the region or part where the problems with sound occur. Thus, the sound analyses stage continues by analysing the effect of the assembly parts of the product on the product sound. This is carried out by disassembly of the product in a by step-by-step fashion and recording at each stage of disassembly until the last sound-producing component is left. Again, acoustical and psychoacoustic analyses are required for each recorded sound. This is a crucial stage in product sound design that aims at determining which existing component of the product is problematic and can be replaced.

As exemplified above, the sound analysis stage is based on many iterative processes that involve observations and analyses into human-product interaction within context, the acoustical content of the sound, and physical construction of the product. Such analyses lead to understanding the conceptual and functional role of sound in human-product interaction.

4.2. Stage 2: Conceptualization

Once the conceptual and functional problems with product sounds are identified during the sound analysis phase, designers can proceed with conceptualizing the to-be-designed new product sounds. The conceptualizing should incorporate the desired product experience (as defined in the product brief) as a reference but focus on the sound-specific relevance to the desired experience. For example, if a shaver is being designed to be *sporty*, the sound does not necessarily have to refer to this concept directly. Semantic associations (i.e., sub-concepts) of *sporty* (e.g., *powerful, dynamic, energetic*) applied on the shaver sound would be also satisfactory as a contribution to the overall product experience.

Therefore, at this stage, it is important first to define the semantic associations of the desired product experience in order to determine what underlying concept could be taken further for sound design. Such conceptual analysis can be made with the help of a couple of methods (Özcan & Sonneveld, 2010). Mindmapping, bodily explorations, and acting out are complimentary methods that help to deconstruct the meaning of a desired experience. With bodily explorations, designers try to put themselves in a, e.g., *sporty* mood and determine situations when one feels sporty (e.g., jogging, playing tennis). They internally observe what happens in their body if they are sporty and further check their emotional state to determine how pleasant, aroused, or powerful they feel. With acting out, designers physically act out, e.g., *sporty* by moving their bodily parts, vocalizing sounds accordingly, and interacting with other objects. This method is important to determine the physical and temporal properties of the desired experience. Once such explorations into meaning deconstruction are complete, designers can summarize their experiences with the help of a mind map (a.k.a. knowledge map). The purpose of the mindmap is to systematically unravel the meaning of a desired experience, which is an abstract term, and relate it to physical properties of objects/interactions/sounds, which are concrete entities. Furthermore, mindmaps often help designers to determine metaphors which may be useful for the application of the concept. As a result, a concept supporting the desired product experience can be taken further for sound sketching.

Once a concept is selected, a next step is to audiolize this concept with sound sketching. The ultimate goal of sound sketching is to find auditory links that may underlie the selected concept (and the desired experience, directly or indirectly). Sound sketching can be done via tinkering with objects, vocalizations of sounds, and/or using a sound sketching tool (e.g., PSST! Product Sound Sketching Tool). With tinkering, designers are encouraged to find objects that can express the desired auditory expression when in interaction with other objects. It is important here how designers tackle the objects, with what actions and movements. Tinkering is all about creating sounds with ordinary daily objects. With vocalization, designers can vocally imitate the sound with auditory expressions of the desired experience. For example, having learnt during the prior bodily/physical explorations that a *sporty* sound should be *energetic, dynamic, determined*, etc., designers can vocalize an engine sound with such auditory expressions. Finally, if designers have access, they can sketch sounds with a specially designed interactive tool such as PSST! (Jansen, Özcan, & van Egmond, 2011). PSST! allows designers to create digital sounds with previously recorded samples by manipulating the timbre, sound intensity, and pitch. PSST! is more suitable for consequential sounds.

The conceptualization phase is complete once the desired auditory expression has been determined. The sound sketches can be further used as a guide for the prototyping of the product with the desired auditory expression.

4.3. Stage 3: Embodiment

In the design and construction of the products, the embodiment phase is the first moment when designers encounter sounds emitting from the newly designed product. The embodiment phase for sound design concerns the physical product parts that need to be altered/replaced in order to create the desired auditory experience. Therefore, the problematic parts encountered in the analysis stage will be tackled at the embodiment stage. One activity that is essential to this stage is the prototyping. Designers need to partially prototype the product in order to observe the occurring sound and verify its fit with the desired auditory features and experience.

Similar to the sound analyses stage, each occurring sound needs to be acoustically analysed. The same methods of sound recording and analysis such as used in the analyse phase can be adopted here. However, the observations and conclusions should be tackled around the desired auditory experience.

Tools and methods used for the embodiment design of sounds depend on the type of sound. Intentional sound design and application require more digital techniques to construct the sound and consequential sound design and application would require more analogue techniques to construct the product, hence the sound.

4.3.1. Intentional sounds

Intentional sounds are by nature music-like sounds, thus they can be created from scratch with the help of a musical instrument or a computer with proper sound editing tools (e.g., Garage band, Audacity). Timbre, temporal structure, and length are some factors that need to be considered when designing intentional sounds. The intentional sounds are already described in chapter 3.

4.3.2. Consequential sounds

For example, if a food chopper is producing an unwanted fluctuating sound and it has been found that the mill that turns the blade was found vertically tilted due to bad assembly; then, a better construction that stabilizes the mill could be proposed. In another example, the working principles of a coffee machine could be altered by… in order to create the feeling of efficiency and comfort. Furthermore, once the main assembly of the product is finished and a rough sound can be produced, it is possible that old-fashioned techniques of noise closures and dampening could be employed before the casing is designed and assembled.

The embodiment design phase is complete once the guidelines for the final prototype are achieved. It should be kept in mind that the product sound occurring at the prototyping stage may be different to the sound of the final product. Thus, the embodiment sound design phase consists of iterative stages of creating sounding models, (dis)assembling, and testing with the aim of achieving the desired experience with the final product. The tests involved here range from acoustical measurement and analysis of the sound via a computer to see whether the product sound fits the technical requirements, or cognitive evaluation of sound with potential users to ensure that the occurring sound semantically fits the desired experience. Moreover, with the sounding models, desired interaction with the product can be enabled and observed. This could be done with the help of potential users acting out towards the product and the design team, enabling the interaction with the wizard-of-Oz techniques.

4.4. Stage 4: Detailing

In the detailing phase, fine tuning of the product sound takes place. At this stage, the final prototype is built and the product to-be-produced takes its final shape. A more realistic sound is expected as an outcome. More extensive user research takes place with semantic differentials and observational studies. Collected data should yield more accurate results and conclusions regarding the desired experience and interaction. It is possible that the occurring sound still needs further adjustments. At the detailing stage, there will be room for further noise closure and dampening activities that roughly concern the outer shell of the product. At the end of detailing, the product should be ready for manufacturing.

5. Product sound designer

Sound design activities exemplified above are multi-disciplinary by nature and relate to three indispensable disciplines: *acoustics, engineering,* and *psychology.* Each of these disciplines contributes equally to the sound design process and a sound designer needs to have insights into each of them. Figure 4 demonstrates how knowledge from these disciplines feeds the sound design process. In the following paragraphs we will explain the individual contribution of these different fields of expertise and create the profile of a sound designer.

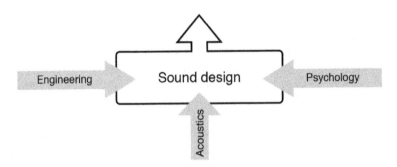

Figure 4. Main disciplines contributing to product sound design activity (adapted from: Özcan & van Egmond (2008)).

5.1. Acoustics

Acoustics is the science that tackles sound phenomena. The field of acoustics is concerned with basic physical principles related to sound propagation and mathematical and physical models of sound measurement. Therefore, the topics of interest for the field of acoustics are the medium in and through which sound travels, reflecting and vibrating surfaces, speed of sound, and other physical characteristics of sound such as sound pressure, wavelength and frequency.

Sound is a result of the energy release caused by objects in action. Although the physical quality of the sound is determined by the sound source and action, *acoustics* does not necessarily investigate the source per se. The physical properties of the source (e.g., the interacting materials, weight, size, and geometry of the objects) are of interest for acousticians. Furthermore, sound propagates over time because it is the result of time-dependent dynamic events. That is, the physical character (i.e., spectral-temporal composition) of a sound changes over time depending on the type of actions and sound sources. For example, a piano produces a harmonically and temporally structured sound. A lady epilator produces a noisy sound because it contains multiple sound-producing events, each creating different harmonic partials and occurring at different time frames, causing temporal irregularity.

It is essential to understand the acoustic nature of the sound event when designing product sounds. Acoustic analysis of the sound can be first done during the problem analysis phase and can recursively occur until the problem has been defined. The field of acoustics provides tools and methods to analyse and simulate sound. Basic terms used for sound characteristics comprise of 'frequency' (variation rate in the air pressure), 'decibels' (sound intensity), and 'amplitude' (sound pressure). A spectrogram visualizes the frequency content of a sound and the intensity variations in time. Furthermore, a sound wave represents the temporal tendency of sound propagation and the sound pressure over time. It is possible to visually analyse the spectral-temporal composition of a sound event and precisely pinpoint the acoustical consequences of certain events. Moreover, various sound modelling techniques have been developed in the field of acoustics. Simulating sounding objects that are perceptually convincing has been possible thanks to the available computer technology (Cook, 2002; Pedersini, Sarti,

& Tubara, 2000; Petrausch, Escolano, & Rabenstein, 2005; Rocchesso, Bresin, & Fernstrom, 2003). Furthermore, sound simulation can also be necessary to test upfront the perceptual effects of the desired sound.

5.2. Engineering

Engineering is the discipline through which abstract scientific knowledge takes on an applied nature. For the design of product sounds, three main branches of engineering provide knowledge: mechanical engineering, electric-electronics engineering, and material engineering. These relevant fields deal with sound indirectly and rather focus on manipulative (i.e., constructible) aspects of products. Various product parts, mechanisms, lay-out, materials, interactions, and working principles can all be engineered depending on the design requirements of the product and its sound.

In product engineering, functionality of the product should be the main focus. Thus, suggested alterations for the improvement of the product sound can only be carried out if the functionality of the product or product parts are kept intact. Engineers should have satisfactory knowledge on physics and mathematics, and they are able to calculate the energy release as sound or as vibration. Furthermore, the discipline of engineering provides various tools and methods to embody conceptual ideas and solutions to problems. Engineers and designers are well-supported on modelling, testing, and prototyping (Cross, 2000; Hubka & Eder, 1988; Roozenburg & Eekels, 1995). Similar tools and methods could be used for implementing product sounds as well.

5.3. Psychology

Sound design is not limited to finding technical solutions for a problem. The aforementioned disciplines deal with the physical aspect of sound and the object causing the sound (i.e., product). However, product sounds, just like other environmental sounds, have psychological correlates which may be on a semantic level or an emotional level (von Bismarck, 1974; Kendall & Carterette, 1995; van Egmond, 2004).

Listeners main reaction to any sound is to interpret it with their vocabulary of previous events. Such interpretations often refer to the source of the sound and the action causing the sound, such as a hairdryer blowing air Marcell, Borella, Greene, Kerr, & Rogers, 2000). Listeners are able to follow the changes in the spectral-temporal structure of the sound and perceive it as auditory events or sometimes as auditory objects (Kubovy & van Valkenburg, 2004; Yost, 1990). In the absence of image, just by hearing listeners can describe the material, size, and shape of the sound (Hermes, 1998; Lakatos, McAdams, & Causse, 1997).

For product sounds the conceptual network consists of associations on different levels (Özcan van Egmond, 2012). Source and action descriptions occur the most, followed by locations in which products are used the most (e.g., bathroom, kitchen), basic emotions (e.g., pleasant-unpleasant), psychoacoustical judgments (e.g., sharp, loud, rough). In addition, source properties can also be identified (e.g., interacting materials or sizes of the products). Furthermore, product sound descriptions could also refer to rather abstract concepts such as hygiene (for the sound of washing machine), wake-up call (for the sound of the alarm clock), and danger (for a warning buzzer).

These conceptual associations of sound indicate that a fit of the sound to the product or with the environment in which the sound occurs is judged. Therefore, a design team cannot overlook the cognitive and emotional consequences of the sound. In various stages of design, user input needs to be carefully considered. Therefore, questionnaires that are aimed at measuring the psychological and cognitive effect of sound could be used.

5.4. Hybrid disciplines: Psycho-acoustics and musicology

Above we discussed the major disciplines contributing to sound design. However, some hybrid disciplines also contribute such as psycho-acoustics and musicology. The field of psychoaoustics deals with the basic psychological reactions to the acoustic event. Sharpness (high frequency content), roughness (fluctuation speed of the frequency and amplitude modulation), loudness (sound intensity), and tonalness (amount of noise in a sound) are the main parameters used to observe the psycho-acoustical reaction of listeners. Although these parameters are supposed to be subjective, a general conclusion has been made in the past regarding the threshold and limits of human sensation to sounds. Therefore, psycho-acoustical algorithms have been presented to measure the above-mentioned perceived characters of sound (Zwicker & Fastl, 1990). These algorithms are used to measure the sounds perceptual quality and predict listeners tolerance to sounds. Thus, they are predictive of sensory pleasantness or unpleasantness.

Designers can design alarm-like synthesized sounds if they have knowledge and practical experience in the field of musicology, as composing music that requires knowledge on theories about musical structures and compositions and tools to create harmonic and rhythmic sounds.

5.5. Responsibilities of a product designer

A product sound designer should have knowledge and skills on three major disciplines (engineering, acoustics, and psycho-acoustics) and also on hybrid disciplines such as musicology and psycho-acoustics (see Figure 5).

A product sound designer is primarily an engineer that is able to manipulate the product lay-out and is skilful in applying physical and mathematical knowledge in order to analyse and to model the product lay-out while considering the consequences in terms of sound.

However, interpreting the physics of sound per se should also be one of the major roles of such an engineer. Skills in acoustic analyses and ability to simulate sound are necessary. Furthermore, a sound designer should be able to link the structural properties of a sound to its acoustical composition. In addition, musical knowledge on how to compose synthesized sounds is required in the case of the intentional sounds.

Furthermore, the psychological correlates of the product sound should also be considered when an engineer is tackling the physical aspects of sound and the product as a sound source. Ultimately, the product sound designer has the last word when judging whether the sound

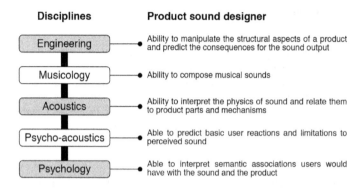

Figure 5. Professional domains of a sound designer (adapted from: Özcan & van Egmond (2008)).

fits the desired experience and the interaction within the context of use. Knowledge on psycho-acoustical analyses is required to predict the first user reactions only to sound. Later, semantic analyses need to be conducted with potential users to make sure the sound design is complete and appropriate to the product.

6. Product sound design course

Product Sound Design is an elective course of the Master of Industrial Design Engineering Education at Delft University of Technology. In product sound design we distinguish two main types of sounds: Intentional sounds and Consequential sounds. The two types of sounds are addressed to the second half year - in the first quarter intentional sound and in the second quarter consequential sound. The students involved are working in project teams of two or three students. The elective consists of a project with few lectures to support the project. The final results should be presented to all course members and stakeholders in a colloquium. The presentation takes approximately 25 minutes with 5 to 10 minutes for questions and discussion. The project is graded on the deliverables: presentation, and report. For the projects, domestic appliances are chosen such as: kids alarm, public public transport card check in and check out, electrical toothbrush, choppers in different versions, shavers, etc..

6.1 Intentional sound project

The intentional sound project approaches the design of these sounds from an interaction perspective. These sounds are synthesized or recorded and are often more musical or speech-like. Therefore, the sounds are created by use of music software. The function of sounds is often to alarm or to provide feedback to users. The project focuses on perception and re-design of these sounds from n interaction point of view . It is essential that these sounds are designed on the bases on the interactions, otherwise improper sounds will result.

6.2 Consequential sound project

The consequential sound project focuses on the sounds radiated by domestic appliances, and are a consequence of their operating and construction. The students will analyse the sounds in the physical, perceptual, and emotional domain and try to relate these findings to the engineering parts of the product. A product should be disassembled and sound recordings will be made of different parts in order to obtain insight in the contribution of these different parts to the sound. The findings resulting from the analyses in the physical, perceptual, and emotional domain are used to redesign new parts or different working principles.

6.3 The study goals

The study goals provide a basis for the self-development of a designer in the field of product sound design. The goals are:

- To be able to implement their findings from the analyses in the physical, perceptual, and emotional domain into an adapted product design,

- To learn how sound is produced in products and experienced by people,

- To learn basic principles of signal analysis (related to sound),

- To learn the effect of tolerances on the performance of the appliances and its sound production,

- To learn the relationship between product quality and sound quality.

The students get 6 ects (European credit transfer system) for this elective course. This means that they have to invest 168 study hours to come to new ideas and realize them in an adapted or innovative product sound design. The valuable results of three years since the elective started were used to develop further on the elective product sound design.

First of all, because the education method of this elective was very successful it will be upheld. The students are working in project teams because of the complexity of the topic. The frontal lectures are limited to project planning and organization, and to an introduction into the basics of product sound design. Sound recording is explained as among which: use of software, lab set-up, and how to record. Most recordings will be carried out in the Audio Lab, but if the project requires recordings can be made at a specific location (e.g. public transport card project). The coaching of the teams is on the initiative of the students which stimulates them in their search for creative and innovative solutions. At specific moments during the project, teams have to explain the project progress. During these moments coaches discuss the progress and results and give advice to go into a certain direction when necessary The results of a project are presented in a colloquium and a written report.

6.4. The case: Toothbrush

We use a student project of a toothbrush as an example. The team measured the sound under load, shown in the lab setup in figure 6. This laboratory setup is easily adaptable to record the

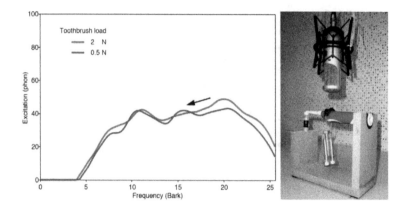

Figure 6. Laboratory setup and Barks analysis for the maximum and minimum load on the toothbrush. The load will be applied by hanging bolts on the toothbrush.

sound of the toothbrush under different loads. A sound level meter is used to obtain the loudness level in decibels. The recordings will be analysed to get insight into the sound effects at different power loads. The brush force for brushing teeth effectively lies between a maximum of 2N and minimum of 0.5 N in normal use.

The maximum load applied to the toothbrush is determined by the operation of the toothbrush at the boundary of the function in this case. The minimum load of the toothbrush is determined by its own weight. Figure 6 shows the graph of 2N and 0.5N loads - the influence of load on the toothbrush can be observed on the Bark scale. A peak is observed at 20 Barks for a load of 0.5N. It moves to a lower frequency domain of 15 Barks when the maximum load is reached.

The disassembly of the toothbrush is carried out in order to analyse the recordings of the parts contributing to the sound. In figure 7 the inside organization of the toothbrush is shown.

Disassembling of the toothbrush from complete product (situation black) to only the electromotor (situation brown) the recordings in barks in figure 8. The different graphs of disassembling the product are given in different colours. With a decrease of number of parts (disassemble step after step), the sound will gradually cut down.

The main axel (situation blue) is the main cause in the irritating rattling noise. When removed (situation green), there is a big decrease in the peak around 20 barks and a lowering of 6dB in volume. In the last stage of disassembly (situation brown), only the motor is active, that results in a sound of 45 dB. Gearing parts are assembled on the motor; this will increase the resistance that contributes to a louder sound, especially in the lower-frequency domain.

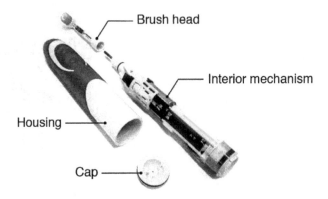

Figure 7. Inside image of the toothbrush.

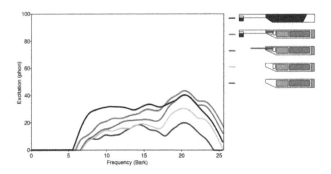

Figure 8. Barks analysis on the complete disassembly of the toothbrush.

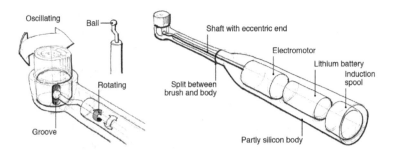

Figure 9. Sketches of the redesign - in this case the working principle is changed.

In final stage, hand sketching is used to express the solution by means of the working principle. Sketching is a handy tool for the designer to visualize the working principle, product ideas, and parts quickly on paper. The sketches show how parts may be produced and assembled. However, implementing it in a product is often not feasible, therefore the intended sound cannot be measured. The toothbrush changes are based on sketching because making a prototype with rapid prototyping could bring you far away from the final solution because the replacement material has never the same sound property. In figure 9 sketches of a redesign are shown.

7. Discussion and conclusion

The two types of product sound need their own design process. Consequential sounds are the result of the product layout. However, the component choice, shape, material and manufacturing are the main parameters that determine the consequential sound. For a new innovative product, sound recordings of different components will be made and mastered into a future product sound (Van Egmond, 2008). In this situation, the product sound will rely on the experience of the product sound designer . In future, this experience should be replaced by a theoretical framework based on research on the following parameters: material, accuracy of parts, the tolerance of parts, how the parts are connected, power transport, size, geometric, speed, and assembly tolerances.

Although consequential sounds are restricted in degrees of freedom by the design and embodiment of the product, the design of intentional sounds has an unlimited number of degrees of freedom due to the fact that they can be designed from scratch. This is one of the reasons that many feedback and alarm signals are badly designed, because no limiting constraints are imposed. If one considers the design of intentional sounds as a form of interaction design, the interaction can impose the constraints on the design of the sounds. Consequently, the sounds will "fit" their function better. Knowledge from interaction design, psychoacoustics, audio engineering and music perception will form the theoretical basis of the design of these sounds.

It can be concluded that Product Sound Design should be a discipline within the domain design. Especially, the physical and psychological aspects should be mapped onto each other. The product sound designer has to learn from a variety of disciplines, from design to engineering and from acoustics to music perception. The course in product sound design is a good basis for further self-development of young designers. It enables opportunities for students under supervision of their lecturers to develop a systematic approach for product sound design. Hopefully, this will lead to more knowledge and appreciation of the way sound contributes to the overall product experience.

Author details

Lau Langeveld, René van Egmond, Reinier Jansen and Elif Özcan

Delft University of Technology, The Netherlands

References

[1] Blauert, J, & Jekosch, U. (1997). *Sound-quality evaluation- a multi-layered problem,* Acustica/acta acustica 83 (1997) same issue.

[2] Cook, P. R. (2002). *Real sound synthesis for interactive applications.* Natick, MA: Peters.

[3] Cross, N. (2000). *Engineering design methods: Strategies for product design* (third edition). Chichester: John Wiley and Sons Ltd.

[4] Edworthy, J, Hellier, E, & Hards, R. (1995). The semantic associations of acoustic parameters commonly used in the design of auditory information and warning signals. *Ergonomics* 38(11), , 2341-2361.

[5] Farnell, A. (2010). *Designing Sound,* MIT Press, Cambridge Massachusetts, USA

[6] Freudenthal, A, Melles, M, Pijl, V, Bouwman, A, & Stappers, P. J. (2005). A Contextual Vision on Alarms in the Intensive Care Unit. Human Factors in Design Safety, and Management, Maastricht, The Netherlands, Shaker Publishing.

[7] Gaver, W. W. (1993). What in the world do we hear? An ecological approach to auditory source perception. *Ecological Psychology, 5*(1), , 1-29.

[8] Hearst, M. A, Albers, M. C, Barrass, S, Brewster, S, & Mynatt, E. D. (1997). Dissonance on audio interfaces." *IEEE Expert-Intelligent Systems & Their Applications* 12(5), , 10-16.

[9] Hermes, D. J. (1998). *Auditory material perception* (Annual Progress Report IPO.(33)

[10] *Heufler, G. (2004). Design basics, from Ideas to Products,Niggli Verlag AG, Sulgen, Zürich.*

[11] Hoggan, E, Raisamo, R, & Brewster, S. A. (2009). Mapping information to audio and tactile icons. Proceedings of the 2009 international conference on Multimodal interfaces. Cambridge, Massachusetts, USA, ACM: , 327-334.

[12] Hubka, V, & Eder, W. E. (1988). *Theory of technical systems: A total concept theory for engineering design.* Berlin: Springer.

[13] Jansen, R. J, Özcan, E, & Van Egmond, R. (2010). PSST! *Product Sound Sketching Tool.* AES: Journal of the Audio Engineering Society, 59(6), 396-403.

[14] Kendall, R. A, & Carterette, E. C. (1993). Verbal attributes of simultaneous wind instrument timbres: I. von Bismarck's adjectives. *Music Perception, 10*(4), 445- 468.

[15] Kubovy, M, & Van Valkenburg, D. (2001). Auditory and visual objects. *Cognition, 80*(1-2):97-126.

[16] Marcell, M. E, Borella, D, Greene, M, Kerr, E, & Rogers, S. (2000). Confrontation naming of environmental sounds. *Journal of Clinical and Experimental Neuropsychology, 22*(6), 830-864.

[17] Özcan, E, & Van Egmond, R. (2008). Product Sound Design: An Inter-Disciplinary Approach?, Design Research Society Biennial Conference (DRS2008), Sheffield, UK.

[18] Özcan, E, & Van Egmond, R. (2012). *Basic semantics of product sounds*. The International Journal of Design, in press for August, 2012.

[19] Özcan, E, & Sonneveld, M. (2009). Embodied explorations of sound and touch in conceptual designing. Proceedings of the 5th International Workshop on Design & Semantics of Form and Movement: Taiwan: DeSForm.

[20] Pedersini, F, Sarti, A, & Tubara, S. (2000). Object-based sound synthesis for virtual environments- Using musical acoustics. *IEEE Signal Processing Magazine, 17*(6), 37-51.

[21] Petrausch, S, Escolano, J, & Rabenstein, R. (2005). *A General Approach to Block-based Physical Modelling with Mixed Modelling Strategies for Digital Sound Synthesis*. Proceedings of the IEEE International Conference on Acoustics, Speech, and Signal Processing, Pennsylvania, USA.

[22] Rocchesso, D, Bresin, R, & Fernstrom, M. (2003). Sounding objects. *IEEE Multimedia, 10*(2), 42-52.

[23] Rocchesso, D, & Polotti, P. Delle Monache, S. ((2009). Designing Continuous Sonic Interaction." International Journal of Design , 3(3), 13-25.

[24] Roozenburg, N. F. M, & Eekels, J. (1995). *Product Design, Fundamentals and Methods*, Wiley, Chichester, UK.

[25] Rothbart, H. A. (1964). *Mechanical Design and Systems Handbook*, McGraw-Hill, New York, USA

[26] Sanderson, P. M, Liu, D, & Jenkins, S. A. (2009). Auditory displays in anesthesiology." Current Opinion in Anaesthesiology 22(6): , 788-795.

[27] Verheij, J. W. (1992). *Basics Soundless Engineering*, (Lecture book), Faculty Mechanical Engineering, Technical University of Eindhoven, Eindhoven, Netherlands.

[28] Von Bismarck, G. (1974). Timbre of steady sounds: A factorial investigation of its verbal attributes. *Acustica*, , 30, 146-159.

[29] Yost, W. A. (1991). Auditory Image Perception and Analysis- the Basis for Hearing. *Hearing Research, 56*(1-2), 8-18.

[30] Zwicker, E, & Fastl, H. (1990). *Psychoacoustics: Facts and models*. Berlin, Heidelberg: Springer.

Industrial Design Perspectives

Chapter 5

The Design of Product Instructions

Dian Li, Tom Cassidy and David Bromilow

Additional information is available at the end of the chapter

1. Introduction

This chapter follows the chapter 3 (Product Instructions in the Digital Age) in Industrial Design - New Frontiers (2011) and describes the design process used in designing effective hard copy and interactive digital instructions for a selected product, which was a photo table. In this research, the printed and multimedia instructions had to be planned, using exactly the same contents, text and illustrations, in different formats to find out the impact of different media on instructions. The authors referred to literature, related research and design examples then developed the product instructions and developed a practical design process for designing product instructions.

2. Suggested design process

Pettersson (2002) suggests a design process for instructional messages, which comes with six steps: 1) Analyse requirements; 2) Plan contents; 3) Design language of messages to communicate; 4) Deliver and present messages; 5) Testing; 6) Refine designs.

This paper gives an overview of what to do when dealing with instruction design but further details are not explained. Two aspects of these steps, the design of language and testing are also recognised as important by other researchers. Sherman & Craig (2003) used case studies to understand the design of user documentations and instructions. Their study involved both the language communication and evaluating/ testing of documentations; the overall design processes were not mentioned. Similarly, ISO/IEC GUIDE 37(1995) suggested two types of assessing methods for general user instructions: desk research and user testing. Again it did not make suggestions on the design process of general instructions.

Figure 1. Examination of the product

3. The actual design process

In this research, the authors referred to suggestions on instructional message design and produced two sets of product instructions: printed instructions and multimedia instructions. They both used exactly the same contents, text and illustrations. The shared information was vital to ensure equality comparison of the effectiveness between both sets of instructions. To guarantee its quality, the product was examined first, then the user profile was reviewed and tasks were analysed. Contents were then decided and written according to the recommended communication rules, followed by the design of visuals and the production of instructions. This research focused on the instructions which accompany the product, therefore instructions on packaging and the product itself were not studied.

3.1. Examination of the product

In the examination, all parts of the product were inspected and measured (Figure 1). Although this research focused on the instructions which accompany the product, labels had to be added in term of using the accompanying guides properly (Figure 2). Further, experiments on using the product, for example installing and operating were performed (Figure 3). The process was observed and recorded by a camera.

3.2. User profile

After the examination of the product, its user profile was created. This process ensured that all instructions would satisfy the product's appropriate user groups and meet their special requirements if there were any.

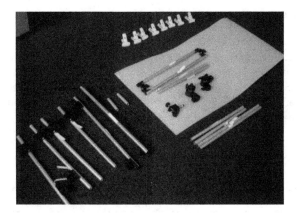

Figure 2. Labels were added onto product parts.

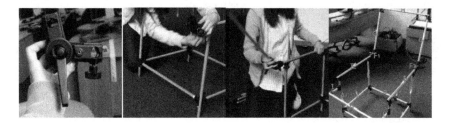

Figure 3. Examine the installing and operating procedures.

The chosen product was designed for photography practitioners, including both experts and novices. Photography experts might have more chances to expose themselves to similar products before but it was a simple product therefore separate instructions for experienced and new users were not necessary. The redesigned instructions should require the minimum knowledge from users and be able to be used by people from all experience levels. It was important that users should be physically capable of carrying out the required actions because the product needed installation and adjustments for use (Table 1):

Product name: Photo table			**Model**: ST-0613T	
User knowledge and experience:				
Reading level:	low	medium	high	<u>all levels</u>
Product experience:	experienced		novice	<u>both</u>
User of similar products:	many	some	little or none	<u>all</u>
Physical characteristics:	not mobility impaired			

Table 1. User profile of photo table ST-0613T.

3.3. Possible tasks

Users mainly need instructions either for learning to use a product, for troubleshooting or for both. In this experiment, the main purpose of the instructions should be providing sufficient information so that users could set the photo table up to use safely. The possible tasks, sub tasks, actions and their ideal solutions, acceptable/ alternative solutions even possible errors were listed and analysed (Table 2). This detailed task analysis was a foundation for planning the contents of instructions.

	Ideal solution	Acceptable solution	Possible errors
Finding general information			
Locating information	Locate information quickly	Locate information slowly	Cannot locate information
Reading information	Read information correctly		Read information incorrectly
Preparing the product for assembly			
Unpacking	Unpack all parts		Miss parts
Matching every part to instructions	Check and match all parts before assembling	Check and match parts during assembling	Mismatch parts
Assembling the product			

→Making the frame base
•Preparing A(L)", "A(R)", "B"× 2;
  -finding A(L)
  -positioning A(L)
  -finding A(R)
  -positioning A(R)
  -finding "B" × 2
  -positioning "B" × 2
•Connecting "C"× 2 to "A"legs and "B"legs;
  -finding "C" × 2
  -positioning "C" × 2
  -connecting first "C" and A(L)
  -tighten the knobs
  -connecting first "C" and "B"
  -tighten the knobs
  -connecting second "C" and A(R)
  -tighten the knobs
  -connecting second "C" and "B"
  -tighten the knobs
	Making the correct frame base in designed order
	•Right parts face correct directions;
	
	•Right parts are fixed in the right joints;
	Parts are safely fitted into each other;
	Knobs are tightened;
		Making a correct frame base in any order	**Can not make a correct frame base**
			•Use wrong parts;
			Right parts face wrong directions;
			
			•Connect wrong legs;
			Connect right legs in wrong joints;
			Parts do not sit into a secure position;
			Knobs are not tightened properly;

Table 2. Examples of possible tasks for using the photo table ST-0613T and the required actions.

3.4. Planning contents

Having the users' needs in mind, the contents of the instructions were then redesigned. Compliance checklists from the standards ISO/IEC GUIDE 37(1995) and the BS EN 62079:2001/ IEC 62079 (2001) were used as references. The contents applicable for the chosen contents were organised to cover five key elements: product identification, product specification, preparing the product, operating instructions and health and safety information (Table 3). This particular product is very simple and does not require trouble-shooting instructions.

Product identification	Brand and type	Falcon Eyes ST-0613T Photo Table
	No. of model	ST-0613T
	Date of publication of the handbook	The product instructions were redesigned on August 2009
	Producer/supplier, Distributer	Manufacturer: Falcon Eyes Limited BENEL BV; Nabliudatelnyje Pribory Ltd.
	Address, etc. of producer/ service agency	Contact details of both manufacturer and distributers
Product specification	Dimensions	130x60cm (back height is 60cm) (words combined with illustrations/animations)
Preparing the product	Unpacking	Parts list (in both words and illustration)
	Installation and assembly	Assembly Instructions (words combined with illustrations/animations)
Operating the instructions	Complete for correct intended use	Operation Instructions (words combined with illustrations/animations)
Safety and health information	Warnings	Not suitable for children It is not a chair Keep the instructions for future reference

Table 3. Contents of the revised product instruction.

3.5. Planning the communication of written instructions

To communicate successfully, the product instructions were written in a clear style and active voice. The instructions were in short sentences; each sentence gave one command and the commands were direct (Table 4). The written instructions followed the communication process and offered users a continuously improved understanding. Terms, information and communication styles were consistent in all parts of the product information.

Before revision	After revision
	Use "C" to connect "A"x2 and "Bx2";
Connect 2 "A" and 2 "B" onto "C" tubes as per photo.	Tighten the knobs.
Then, put on the 2 "D" onto it.	Use one "D" to connect A(L)and A(R), another "D" to connect B and "B".

Table 4. An example of rewritten instructions.

3.6. Designing the visuals

The main challenge for the visual design in this research was to be clear and make sense. Luckily many studies recommended visual principles for making images for instructional use. For example, Szlichcinski (1984) found factors that affect the comprehension of pictographic instructions. Heiser et al. (2003) recommended some cognitive design principles for visualizations. Also, a research focused on effective step-by-step assembly instructions was carried out (Agrawala et al., 2003). Schumacher (2007) reviewed other studies and did more up to date research on pictorial assembly instructions. Together, these studies made clear suggestion for making illustrations, especially in assembly instructions.

The authors used these findings, created a huge amount of illustrations to describe the product, demonstrate the product parts and help to explain the assembling and operating process. To ensure the accuracy of details in illustrations, 3D models of the product were created in Autodesk Maya using the exact proportions (Figure 4).

Figure 4. models reflected the real product.

All models were kept simple and basic. The details of product parts were controlled to the minimum level for easy recognition and 3D models were then rendered as vector images (Figure 5).

Bitmap image X   **Vector image ✔**

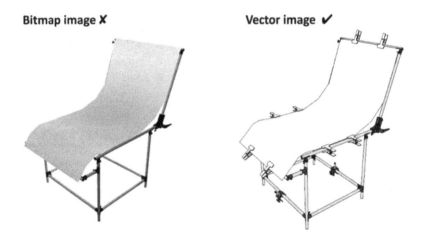

Figure 5. Rendered image examples of the product.

3.7. Colours and text size

The majority of information was designed in greyscale. An orange colour was used to highlight numbers and some icons (Figure 6). All information was guaranteed to be recognisable when printed out in black and white. Text was designed in black on a white ground to ensure a high colour contrast on both print and digital media. Font sizes varied between titles and body text and they were clearly legible.

2 Assembly Instructions

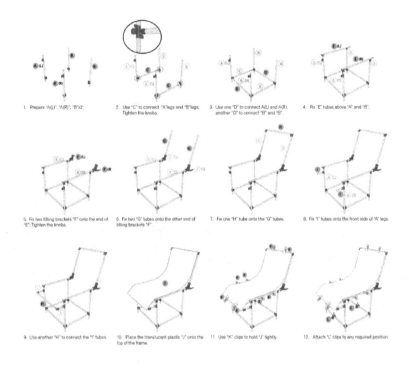

1. Prepare "A(L)", "A(R)", "B"x2;

2. Use "C" to connect "A"legs and "B"legs; Tighten the knobs.

3. Use one "D" to connect A(L) and A(R), another "D" to connect "B" and "B".

4. Fix "E" tubes above "A" and "B".

5. Fix two tilting brackets "F" onto the end of "E"; Tighten the knobs.

6. Fix two "G" tubes onto the other end of tilting brackets "F".

7. Fix one "H" tube onto the "G" tubes.

8. Fix "I" tubes onto the front side of "A" legs.

9. Use another "H" to connect the "I" tubes.

10. Place the translucent plastic "J" onto the top of the frame.

11. Use "K" clips to hold "J" tightly.

12. Attach "L" clips to any required position.

3 Operation Instructions

Turn Tilting Brackets "F" to adjust angle of the panel.

Turn Tilting Brackets "F" to adjust angle of the panel.

Figure 6. Colours used in prototypes.

3.8. Numbering and highlights

The parts, tasks and steps were organised by numbers and their appearances were designed to be coherent. All essential information was either enlarged or emphasized by colours. For example, the warning message was highlighted with a yellow colour to raise attention (Figure 7).

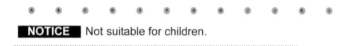

NOTICE Not suitable for children.

Figure 7. An example of highlighted warning message.

3.9. Producing the printed and multimedia instructions

The printed instructions followed the tradition of using text and images to make sense. It was printed on double sides of durable paper and the layout was carefully planned to ensure readability.

Interactive multimedia production instructions were created based on the same contents as that used in the printed instructions. Instead of a linear presentation, information was delivered by re-structured contents. They were re-categorised and designed to be interactive so that users should be able to search and locate information easily. Main visual elements, for example, the written instructions and illustrations remained the same as they were in the printed instructions. Other media like sound and animations also have been integrated together with images and text to provide more effective guidance for users.

3.10. Evaluation based on standards

For the evaluation, the assessment guide provided by international standard, ISO/IEC GUIDE 37 (1995) was referred to (Table 6). Both printed and multimedia instructions satisfied all required and applicable requirements. For the multimedia instructions, the evaluation criteria on their interactivity, user experience etc. are not given in the standard.

Very good	Good	Average, just acceptable	Poor	Very poor	Not applicable/not necessary
++	+	#	-	--	0

Table 5. Add Caption.

3.11. Diagnostic testing

The product instructions were tested to identify any problem in use. Participants were asked to use given instructions to perform a set of tasks. During the test, the participant was required

Items to be checked	Evaluation (++/+/#/-/--/0)	Comments
Legibility		
➔On-product information		
◆Type size depending on reading distance	+	All parts are labeled to enable easy
◆Brightness contrast (needs to be more than 70%)	+	assembling. Although this is not part
◆Instructions incorporated in material of product	+	of this investigation.
➔Handbooks, manuals, leaflets		
◆paper quality	+	
◆type size	+	Black type on white high quality
◆line spacing	+	paper, legible. Different font size for
◆use of different typeface, type size etc.	++	headings.
◆captions easy to read	++	Clear. Orange for highlights. Yellow
◆brightness contrast (needs to be more than 70%)	+	for warning messages. Page layout
◆use of colours	+	is considered
◆general impression of the page is balanced and uncluttered	+	

Table 6. Examples from the evaluative checklist for the re-designed printed instructions.

to "think aloud", speak out her/his thoughts. The participants' actions and voice was observed and recorded for analysis. The results were evaluated by dialogue analysis and error observations.

3.12. Dialogue analysis

The categories for the dialogue contents included confusion, statement of problems, decision made, solution found, statements of intention, considering, statements of feeling, comments on the product and activities like reading instructions and actions. They were marked using different colour codes (Table 6). Physical activities like reading instructions and doing things were marked by blue colours in different tones. Negative thinking processes that involved difficulties and problems were represented by warm oranges. Positive judgments towards some decisions and solutions were drawn in greens and pinks were used for the others.

Thinking activities	Physical activities
Confusion	Reading instructions
Statement of problems	Action
Decisions made	
Solutions found	
Statements of intention	
Considering	
Statement of feelings	
Comments on the product	

Table 7. Categories for the recorded dialogue contents.

Timing for each activity category was recorded and wrote down and it was discovered that overall the tasks were performed well with both instructions. Participants were spending most of the time on physical actions; a reasonable amount of time was used for considering and making decisions. The users also had some problems and expressed confusions (Figure 8).

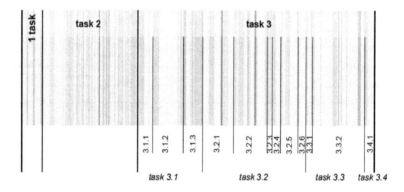

Figure 8. An example of different activities in all tasks presented by colours.

The majority of time had been spent on task 3, assembling the product and this was also the stage where most problems occurred. For example, the time spans for task 3.2.1, task 3.2.2 and task 3.3.2 were much longer than the others and there were negative responses involved in these steps. To clarify details of where the problems were, coloured square icons were used in the following table (Table 8) to indicate confusions and stated problems:

Task No.			Confusion	Stated problems	No reported problem
1					✓
2			◆	■	
3	3.1	3.1.1			✓
		3.1.2			✓
		3.1.3		■	
	3.2	3.2.1	◆		
		3.2.2	◆	■	
		3.2.3		■	
		3.2.4			✓
		3.2.5			✓
		3.2.6			✓
	3.3	3.3.1			✓
		3.3.2		■	

Table 8. An example of the distribution of confusion and stated problems

The negative expressions from the participant like the confusions and problems were listed and analysed (Table 9). These subjective expressions from the participants explained what problems they had been feeling. By inspecting the instructions, some of the problems(Task 2 and 3.1.3) were discovered to be misunderstandings and some of the real problems in Task 3.2.2 were solved.

3.13. Error observations

The dialogue analysis showed that participants were confused at some stages, for example when trying to complete Task 3.2.2 and Task 3.3.2, problems and confusions were pointed out. To study real problems of the product instruction more objectively, the user test results were analysed again by reviewing the video, checking times for each task and observing errors during the test (Table 9).

Through the observations, one error was found while the participant was carrying out Task 3.2.2 using the multimedia instructions, when an "F" bracket was fixed onto the opposite side at the beginning. The error was identified and corrected by the user himself after checking against the instructions. This showed that the product instructions explained the operation process but they could be clearer and more effective in terms of preventing misuse of the product.

Sometimes participant pointed out a few problems and the time span for certain tasks was observably longer than others. However, no error was discovered in the process of finishing it.

Overall, both the printed and the multimedia product instructions performed well and no major mistake was discovered. Still, improvements were carried out to reduce users' confusions therefore shorten the time for some tasks, for example task 3.2.2 and task 3.2.2.

3.14. Refine instructions

Findings from the diagnostic tests suggested some major improvements for the instructions. These changes were applied to both the printed and the multimedia instructions. However, in this experiment, after making changes, the instructions still would not fix all discovered problems. This was because of design deficiencies in the product itself and instructions could not and should not compensate for those product inadequacies. Due to the aim of this investigation being focused on the usability of instructions; the minor confusions caused by the design deficiencies were ignored.

4. Conclusion

As stated in 1.1, the current research and study on the design of general product instructions is relatively rare. It is to the authors' surprise that no existing clear suggestion on the instruction creation process has been found and the method of instruction creation is also undefined. Moreover, there are no easy guidelines for designers to follow. The standards on user instruc-

Tasks No.	Thinking activity	Time (s)	Contents	Analysis	Can problems be solved by altering instruction?	Solutions
Task 2	Statement of problems	4	There should be another one... like that.	The participant could not find the part "A(R)". It was found in 4 seconds.	Has been solved	
Task 2	Confusion	5	So, why was that? Have I miss something in it?	The participant could not find one of the "B" parts. It was found in 5 seconds.	Has been solved	
Task 3.1.3	Statement of problems	4	Obviously I have made a mistake.	The participant assumed that "A" pipes were fitted in incorrect positions. However, he proved them to be right after checking instructions.	Has been solved	
Task 3.2.1	Confusion	6	Ah! Why wouldn't I be able to...Where was those two?	The participant was trying to figure out the correct direction of using A pipes and B pipes.	Yes	The illustrations for pipe ends should be further enlarged for easy checking.
Task 3.2.2	Statement of problems	6	Ah, this is left right.	The participant fixed the "F(L)" part in the position for "F(R)" part. The problem was identified by using instructions and on product labels.	Has been solved	
Task 3.2.2	Confusion	8	Em... I would expect that to be...	The participant intended to fix the two brackets "F" in the wrong direction. The problem was identified and solved by using instructions.	Has been solved	
Task 3.2.3	Statement of problems	4	The biggest problem I got was - have to move backwards and forwards.	The participant had to move backwards and forwards to read instructions on screen and take some actions. The instructions were displayed on a portable laptop computer. Therefore this could be avoid by taking the computer and place it closer to the product parts.	No	The tested participant was not familiar with the digital device. The situation varies depend on how much each user is accustomed to different devices. Also, this problem would not occur if the product instructions were displayed on smaller and more portable devices for example smart phones.
Task 3.3.2	Statement of problems	6	These are bit... It doesn't seem to say any particularly. So ...	The participant was trying to find out which side of clip "K" should face up.	Yes	This can be solved by slightly enlarge the illustrations for this part. However, this does not influence the use of the product. Either side of a "K" clip could face up.
Task 3.3.2	Statement of problems	2	It's a bit awkward.	The plastic "J" was left too long at the bottom of the frame and could not be clipped on. This was partly caused by the design deficiencies of the product however can be solved by clarify the assembling order in the product instructions.	Yes	The instructions could explain that the "K" clips on top should be fixed first. Then the plastic "J" should be adjusted to catch the bottom of the frame. Although this cannot fully cover the design deficiencies of the product, it should improve the results.

Table 9. The analysis of confusions and problems in the diagnose test for the multimedia instructions.

Task	Ideal solution	Acceptable solution	Error
3.2.2 Fixing "F" × 2 onto "E" × 2; 3.2.2.1 -finding "F(L)" 3.2.2.2 -finding "F(R)" 3.2.2.3 -positioning "F(L)" 3.2.2.4 -positioning "F(R)" 3.2.2.5 -fixing "F(L)" onto the end of "E(L)" 3.2.2.6 -tighten the knobs 3.2.2.7 -fixing "F(R)" onto the end of "E(R)" 3.2.2.8 -tighten the knobs	◆Right parts are fixed safely; ◆Handles on "F" × 2 face out; ◆Both shorter sides of "F" × 2 are connected to "E" × 2; ◆Knobs are tightened;	◆Handles on "F" × 2 face out; ◆Both longer sides of "F" × 2 are connected to "E" × 2;	◆ One longer side and one shorter side of "F" × 2 are connected to "E" × 2; **(Corrected by the user)**

Table 10. An example of observed errors in the diagnostic testing of the multimedia instruction.

tions are difficult to understand and also very dated. The British standard is relatively new compared to the ISO/ IEC 37. Still it has been out for more than one decade. The guidance on multimedia and digital instructions in these regulations is not sufficient to use. The authors believe that a design process model and an easy guide for producing general product instructions would be very useful and beneficial for instruction designers.

5. Recommendations

To conclude the findings from this research, the authors suggest a design process for instruction planners (Figure 9).

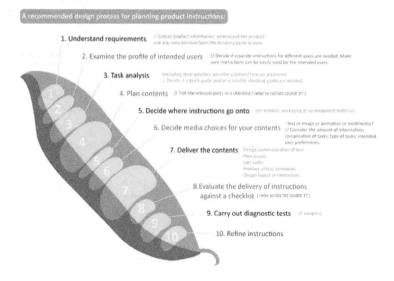

Figure 9. A recommended design process.

This recommended process can be used in combination with the checklist provided by ISO/IEC GUIDE 37 (1995) and the checklist can be updated when the ISO guide is refreshed. It is simple, visual and easy to follow. It should help instruction designers especially those new to planning product instructions. It can also contribute to the development of instruction planning tools.

Author details

Dian Li, Tom Cassidy and David Bromilow

The School of Design at the University of Leeds, UK

References

[1] Agrawala, M.,Doantam, P., Heiser, J., Haymaker, J., Klinger, J., Hanrahan, P., & Tyversky, B. (2003). *Designing Effective Step-By- Step Assembly Instructions.* SIGGRAPH.

[2] British Standards Institution. 2001. BS EN 62079:2001/ IEC 62079:2001. *Preparation of instructions - Structuring, content and presentation.* Milton Keynes: BSI.

[3] Heiser, J., Tyversky, B., Agrawala, M., & Hanrahan, P. 2003. *Cognitive Design Principles for Visualizations: Revealing and Instantiating.* 25th Annual Meeting of the Cognitive Science Society.

[4] International Organisation for Standardisation. ISO/IEC Guide 37:1995. *Instruction for use of products of consumer interest.* Geneva: ISO

[5] Li,D., Cassidy, T. and Bromilow, D.(2011). *Product Instructions in the Digital Age* In: D. A. Coelho (Ed.) 2011. Industrial Design - New Frontiers. Croatia: InTech. Chapter 3(p39-p52).

[6] Pettersson, R. 2002. *Information design: an introduction.* Amsterdam, Philadelphia: John Benjamins Pub. Co

[7] Schumacher, P. 2007. Creating effective illustrations for pictorial assembly instructions. *Information Design Journal*, 15(2), pp. 97-109.

[8] Sherman, W.R and Craig A. B.2003. *Understanding virtual reality: interface, application, and design.* Amsterdam; London : Morgan Kaufmann.

[9] Szlichcinski, C. 1984. Factors affecting the comprehension of pictographic instructions. In R.S. EASTERBY & ZWAGA, H.J.G.(Ed.), *Information Design* (pp.449–466). Cichester: Wiley & Sons.

Chapter 6

Measuring Design Simplicity

Carlos A.M. Duarte

Additional information is available at the end of the chapter

1. Introduction

From John Maeda (2006) it's clearly known that the study of what is Simplicity is central to Design and Engineering. This chapter deals with this, introducing a method to measure Simplicity.

Design and Engineering are today less an act of drawing or designing something but rather the act of designing a program that in itself conceives a diversity of solutions pertaining to the service or function that we intend to draw or design.

This drawing and designing activity may thus be defined by the creation of new materials; genetic manipulation; software and interface conception; formulation of new forms of languages, mostly those of visualization nature; conception (design) of social, political and cultural ideas; generation of new behaviors with growing complexity.

By opposition and in consequence of complexity and also from the intervention of Design and Engineering we all can access better life quality, we can access better technological artifacts and products while allowing its interaction in a simpler way.

Hence this chapter aims the essence of "simplicity" and how it shows up in several existences, whether it may be in Design, whether it may be in Engineering.

The first time ever someone described how the better organization and functionality of systems can be linked to Simplicity was, in 1870, Claude Bernard (Gene, 2007). However Simplicity, as commonly understood, is not an easy thing to describe, much less to comprehend. What is taken as Simplicity is an undetermined number of concepts to explain what supposedly Simplicity is. The result is a great dispersion that cannot afford reasoning. Some have been able to reduce the given conditions for the existence of Simplicity to ten concepts designated as laws (Maeda, 2006). However we cannot consider it as a definition to describe "simplicity" because such multiplicity is just and only descriptive. If we ought to have a classic image of the state of the art for Simplicity is we can remember the first knowledge of the relation between triangle sides, which was an in-equation:

$$a + b \geq c \tag{1}$$

In-equations have an infinite number of solutions. For millenniums solutions were published with possible triangles. The amount of three side set measures could close in a triangle. So it was until Pythagoras simplicity:

$$a^2 + b^2 = c^2 \tag{2}$$

As equations have a finite number of solutions never again was it necessary to write down tables for possible triangles. The comprehension of real space metrics brought up the great growing efficiency to those who design structures. The solution for the problem of knowing if three given measures of line segments would originate a triangle is said to get simplified, and metrics allow simplicity.

Just as in the time when one could only know that the addition of two sides of a triangle was bigger or equal to the length of the third side, to find the solution for simplicity we can also, today, establish tables.

One of the most recent (2006) is that of Maeda. It functions as a synthesis proposal in the rules of Simplicity built in ten laws judged by being different and independent from each other.

This chapter aim is to present a measure for simplicity that will turn it into a parameter central to Design and Engineering.

Until now there are a certain number of rules or laws for Simplicity. The most recent attempt to reduce diversity of notions, Maeda (2006) defines ten laws. But every time we arrive at number ten we cease the main question again. Why ten? Why not eleven or nine? Maeda and his theory of Simplicity suffer from the fact of not having been brought to the Science domain, meaning it has not been made measurable.

Thus this chapter's object is also to clarify Information's concepts, taken as data synonymous, stated arguments and rarely faced as Science parameters, either in Design or Engineering, so thus measurable, though observer dependent and relativistic.

In this context, the chapter is meant to suggest, until designers find out about it, that Simplicity can be measurable and can be so from Fisher's Information Amount (1925) definition. At the same time, the notion will come by that simplicity's measure is also relativistic yet measurable, thus "mathematizable", using concepts initiated by Jacob Bernoulli and those more recently funded in mathematics by Dempster and Schafer (1978).

In 1925, Fischer, upon realizing that just as space and time, information is something we all know until someday someone asks us about it. Yet he also knew that physical language description is based in the fact that, whether or not gathering knowledge of its essence, time and space are measurable. With measure one can add measurement, one can establish comparisons; one can recognize solutions, and the future can be sound.

In 1925, before a set of data, Fisher foresaw the possibility of measurement or calculation of information amount over an unknown variable. A set of data gives an information amount over an incognita that can be measurable or calculated. After Fisher, measuring or calculating the information amount that allows a set of data to acquire uncertainty reducing became a possibility. After Fisher it became possible to know if one is acting upon the incognita, guessing, predicting or foretelling.

The method had been established by Jacob Bernoulli back in 1713. In his "The Art of Conjecture" (*Ars Conjectandi*), he establishes the way to calculate the probability of happening. This probability (*probilitas*) is not the case. It is not a synonymous of *chance*. It is the method of how sure one is about something that is going to happen to the incognita. Bernoulli proposes the method we have selected to present the theory we are about to show, the method of combined argumentation.

We will increasingly use the results from Melo-Pinto (1998) in the recognition technique (the knowing of what the incognita is) and the mathematic method of combining arguments developed by Dempster and Shafer. They both have defined the set for each argument and its weight as a function in the System of Beliefs.

This chapter establishes that for a given System of Beliefs there will be as much simplicity as the lesser the number of data set elements that will be giving the higher amount of information about the incognita.

It will also be concluded that before a same set of data there will be Systems of Beliefs for which there will be higher concentration of information about the incognita.

Hence this chapter is about a measure of Simplicity before a System of Beliefs. Simplicity is as bigger as the smaller the number of elements of the data set that brings the higher quantity of information about the incognita. The relativistic character of Simplicity's measure will be evidenced as well as how to manage it following from the System of Beliefs' properties.

In parallel, it will be established that Design concepts, as well as those of Engineering, have been guiding conception's structuring for Simplicity showing that the concepts expressed in this chapter are not as possible triangles tables were before knowing metrics in Euclid's space, but rather a metric for Simplicity.

This principle may be sustained by several applications; particularly through the Universal Principals of Design. It is also built upon arguments or data, as proposed by Lidwell et al. (2010), who stated that Design essence is one of the meanings of the artifact that surround us and that we effectively use in our daily life. Hence, we will demonstratehow Design Simplicity is also an Engineering parameter.

Following Claude Bernard, who in 1870 discovered physiologic medicine and stated that *the condition for a free and independent life lies in the stability of the internal mean*, which naturally depends on the wellbeing of each individual and is built under his own system of beliefs.

Simplicity is also related to a system of beliefs, and as such it will always be associated to the on-looker that applies it or observes it. Hence one can state that "Simplicity" is relativist, once it's System of Beliefs' are dependable and that variants from the same System of Beliefs can be understood as a School. It will suffice that a set of designers or engineers share the same arguments for each one of them as well as the same weight.

2. Simplicity

The social and economic impact of information systems in society is a reality in part due to the progress in telecommunications in its most recent shapes: internet, television, computing systems and mobile phones. However a new challenge arouses in the digital era, related either to Design, Engineering, or Technological level application, which consubstantiates the need to attain Simplicity, either of contents, either above all of the designing of related interfaces.

Actually Simplicity is also a quality that not only arouses the passionate devotion for product aesthetics and design, but also became a strategic key tool to allow businesses to confront their own complications (Maeda, 2006).

Hence, we can state that Simplicity is something we all long for… For example to refer to the unquestionable commercial success of Apple's iPod – a device that has less functionality than others available in the market with the same function: a digital music reader. People want and prefer products they can use with simplicity.

The first time someone described how things are in fact simplicity-related was, back in 1870, Claude Bernard. But simplicity as usually referred can be everything but simple.

According to the meaning pointed out by Portuguese lexical online dictionary (*http://www.lexi-co.pt/simplicidade/*) simplicity is a feminine substantive that may signify three qualities:

• Quality of what is easy to understand or do: a work's simplicity;

• Luxury absence: to live with simplicity;

• Natural, spontaneity: to speak with simplicity.

In the realm of this chapter thus it will be transmitted to the reader the idea that necessarily individuals who work in science have to speak or write or act mandatorily in a complicate and not complex way, makes no sense when one just wants to communicate.

Today, the requirement is to know how to communicate with simplicity, with clarity, interest and repetition, if thus is necessary, to achieve the main goals to transmit understandably (Kotler, 2000).

Already Priest António Vieira, in one of his most important sermons, *"O Sermão da Sexagési-ma"*, spoken in Lisbon at the Capela Real, back in 1655, were he vindicated clarity, simplicity, syntactic and dialectic rigor, rigor in logic thinking, and what he defined as an oratory art of aesthetics simplicity, advocating that the preacher should opt for a single matter, define it, share it, confirm it by the Holly Readings and with reasoning, the preacher should always use, if possible always with examples and rejecting contradictory arguments, hence he might be able to conclude and persuade...

It is this persuasion capacity that by nature is present in human life through different degrees and combinations that allies simplicity and complexity.

But what is effectively Simplicity? Having also apprehended how really complicated it is to establish a measure of Simplicity, Maeda (2006) was able to describe Simplicity in ten (chapters) laws:

- To reduce – the simpler way to achieve simplicity is by means of a conscious reduction;
- To organize – organization makes a system of many look like a system of few;
- Time – time economy transmits simplicity;
- Learning – knowledge simplifies everything;
- Differences – simplicity and complexity need one another;
- Context – what lies in the simplicity's periphery is definitely non peripheral;
- Emotions – more emotions is better than less;
- Confidence – in simplicity we trust;
- Failure – some things may never be simple;
- Singleness – simplicity is to subtract the obvious and to add meaning.

Then he adds three more components, which are:

- Distancing – more seems like less simply by going far, far away;
- Opening – opening means simplicity;
- Energy – use less, win more.

Still according to Maeda, simplifying a design is harder than making it complicated. The great majority of examples he uses are the result of experiences and problems he underwent.

However his considerations about simplicity in life, in business, in Technology or in Design or Engineering, are not to be considered as having been obtained in the basis of a method from a scientific nature.

As well as we do not consider a science that same method of gathering the catalog of squared angles registration that we referred in the framework of this chapter.

It finds out that, as to the meanings of simplicity, we are allowed an infinite variety of data, and as such the information we have is still short, so being before an in-equation as it has an infinite number of solutions.

In this chapter we will be led to know, as Pythagoras did through his theorem, that there is also a way that allows us to be objective, measuring, on the contrary of the description from

Maeda (2006), one can describe Simplicity as it has been sustained by Fisher (1925), as the first definition of information amount.

In a predictable future, Simplicity is thus doomed to be Design and Engineering's motto, of a whole Industry, as it has been featured in it's own time, the discovery of Pythagoras's Theorem... Shortly, we aim to give to the reader's understanding that simplicity is not something intuitive. In fact, today we possess and have access to great data complication, on the other hand the same data are complicated to process, hence we have very little information, which by analogy leads to conclude that if we are facing an enormous simplicity, a few data set will give a great amount of information about the instrument, or the system, or whatever.

Simplicity, like easiness, hardness, information or complication is a matter of the observer's measuring. It depends on his own system of beliefs.

3. Information and information theory

According to (Drekste, 1999) Information Theory is a branch from probability theory and from statistics mathematics that identifies the amount of Information associated or generated by an event occurrence, or the realization of the state of things, by uncertainty reduction and the elimination of possibilities, presented by the event or by the state of things in question. Relatively to Information "choice" and "amount" are always mentioned and how to measure Information.

Sir Ronald Fisher (1890-1962), a renowned scientist from the 20th century with large contributions to Statistics, Evolutionary Biology and Genetics, introduced in 1925 the concept of Fisher Information, long before the *entropy notion*[1] from Claude E. Shannon (1916-2001) with the maximum probabilities technique and variable analysis.

However, the concept and notion of information is vague and intuitive. When we ask a question, we are requesting data from someone. When we watch a television show or a movie, we are absorbing data. While reading a magazine, or listening to music we know we are dealing with some kind of data. In a general manner, we use, we absorb, we assimilate, we manipulate, we transform, produce and transmit data for as long as our existence.

In 1948, Shannon gave us a more recent definition of Information amount, as the measure of freedom of choice of each one of us when we select a message (data) between all messages of a set.

For Shannon, the content of Information in each message consists only in the quantity of numbers (bits), of zeros and ones leading to the message conveyed. Information can thus be processed as a physical measurable amount.

---

[1]*Shannon's entropy has to do with the fact of the addition of all data, which, according to F. Carvalho Rodrigues is incorrect. We must work with a set of data the smallest and simplest as possible, to maximize the amount of Information to acquire.*

When we want to transform data into Information, we must bear in mind three basic rules (Snodgrass, 2007):

- Organizing data: when confronted with a set of grades these must be numerically ordered;

- Data description: when comparing two or more sets of grades these should be referred in the same scale;

- Interpreting data: upon seeing punctuation tests in a graphic way one must bear in mind that the visual representation is an interpretation in full respect to numerical tests.

Information and knowledge are the result of human action while aggregating data (symbols or facts) in a social or physical scope, out of context, not directly or immediately significant (signs). Data setting in a determined context acquiring meaning and value, being thus designated as Information. Knowledge comes from the successive accumulation of relevant and structured Information capable of action production, partially based in experience. The data transformation in Information and therefore in knowledge requires a cognitive effort in the perception of the structure and of the allowance of a meaning and a value.

Information can thus be grouped at several "levels" from the most basic shape to the most complex one. In these different levels we refer (Bernoulli, 1713; Quinn, 1980; Waibel and Stiefelhagen, 2009) to:

- Context: it is a situation in which something occurs, using any relevant Information to characterize the situation (state description of relevant entities) of an entity that might be a person, a place or an object, to an interaction between a user and an application and including both.

- System of Beliefs: as it will be further explained it refers to the arguments and their own weight. The System of Beliefs is relativist, varying from each observer.

- Context: the situation in which something happens using any relevant Information to characterize the situation (state description of relevant entities) of an entity that might be a person, a place or an object, for an interaction between a user and an application, and including both.

- System of Beliefs: as it will be further explained, it refers to arguments and their relative weight. The System of Beliefs is relativist, varying from the observer.

- Objectives: the goals (or the objectives) refer to what we seek to achieve and when we have results to achieve in mind and no reference on how the results must be achieved. It represents what should be obtained but does not specify how to do it.

A possible structure for Information levels is shown above.

- **Bits and Bytes/Signs** are (SINGH, 2007) the raw material, representing binary Information in the format understood by computers, i.e., zeros and ones that correspond to waves of electric impulses. Zeros represent no current going on and ones represent current going on, like an electric switch. It's the lowest existing level in computers and, for that reason, the one that exists in higher quantity. It is impossible to the observer to distinguish what is represented in it, only finding usefulness when it's found in relevant form. Raw material is no longer considered a level of Information.

• **Symbols** are the alphabet used to order bits, shaping finite strings (words) and infinite strings (messages). Words and messages can be shaped as text or as verbal enunciation, numbers, diagrams and even static or moving (video) images. The occurrence of certain symbols may influence the occurrence of others in several time periods, inducing the amount of Information obtained as each symbol is analyzed.

• **Data** are isolated measures about events. According to (Jones, 2000) data are usually seen as being the most fundamental shape of Information. Usually the meaning "data" means something raw or non-refined that must be "polished" as to transform itself in a finished product. One may be "flooded" with data yet it doesn't mean we have information. An example: People and city's names are usually classified as information, while a serial number one gets at the hospital is "data".

• **Information** is (Jones, 2000) data in a context acquiring semantic value understood as sememes[2]. The acquired information needs to be selected, elaborated and analyzed until it can be used and usually denotes data in a combined shape and in a specific purpose. The terms "data" and "Information" are in general and usually used simultaneously, so a vast set of data hides Information only discovered after data analyzing.

• **Meaning** is Information in a System of Beliefs (Dorbolo, 2003). Beliefs and ideas we have form a system whose parts are interrelated in various shapes. This system is dynamic, i.e., it alters itself as new Information is added, and when it happens all new Information is altered by the system. We all have Systems of Beliefs that condition the way we see and the way we relate with the world.

• **Knowledge** is meaning in several contexts. It's a set of aggregations that surpasses the semantic value of each individual element. It has to do with the way things are made and how they exist. The context (Brezillon, 1998), seen as knowledge, enables necessarily to make the distinction between contextualized knowledge (knowledge used in a certain time) and knowledge context (knowledge that restraints contextualized knowledge). The bigger difficulty we face is that, at a certain time for a follower, contextualized knowledge may become contextual knowledge. There are two kinds of Knowledge (Laudon, 2009):

• **Tacit**: concept (Polanyi, 1974) to situations in which the cognitive/behavior processes are conducted by unconscious cognition. Its the kind of knowledge characterized by being knowledge-practitioner, developed through action, experience, ideas, values, etc. so for that reason shared through conversation and changed experiences, through observation and imitation, not being easily codified, described or reduced to rules.

• **Explicit**: its (Sullivan, 2001) a summary based in direct experience that can be easily articulated or codified through a system of symbols which makes it communicated or diffused. This kind of knowledge is to be found in products, patents, source software code,

---

2 According to Porto Editora Portuguese Language Dictionary, the word "sememas" means the signification unit in a lexeme (lexema), which pertains in a set of semes (semas), the minimal significant component of a word.

etc. Explicit codified knowledge is valuable, raising the capacity of observed and negotiable knowledge, easing communication and learned codification transmitted in rules and thus enduring.

- **Perception** (Frazer and Norman, 1977; Dember et al., 2011) is the meaning filtered by the System of Beliefs. It's the process by which organisms interpret and organize the sensation to build a meaningful experience of the world, generally enhancing processing additional sensorial inputs, the process by which sensorial stimulation is converted in organized experience. The sensation is usually referred to non-processed sense results and, either sensation either perception, in practice, are hard to separate.

- **Meaning** (Frazer and Mackay, 1975) is the meaning understood in terms of objectives. In statistic terms, for very simple models it is possible to find meaning in a numerical equivalence. The test of meaning has the broadest range of applications and requests a model to make the description of an answer as a whole or just a reduced answer.

It is so deductible that the Theory of Information explores the possibility of quantitatively measuring a message's information to utter analysis of its meaning (Carvalho Rodrigues, 1989).

From this we can conclude that before a set of data it is possible to "know" if it has more or less Information over a variable we do not know but want to know.

4. System of beliefs

In the past, guessing was a possible science, having the objective of trying to determine the meaning and causes of events. With science evolution, guessing was abandoned taking into account only predicament and certainty.

All that is known beyond doubt (Bernoulli, 1713), we claim to know or understand. Relatively to all that is left, we only conjecture and opine. To conjecture about something is the same as to measure the probability of something the best way possible, in order to choose the best option to our judgments and actions.

Randomness is not part of our knowledge but the object's property, rendering it impossible to make predicaments about it. The probability is a measure of how certain we are and it's achieved with a combination of arguments. An argument may be defined as a thought to prove or refute a given question.

When an argument fits with mathematics, one can make predictions. When an argument is an image, one can forecast as in "an image is worth a thousand words" as being objective. When one has both arguments (image plus mathematics) one can acquire certainty. Each argument must have a weight and the set of arguments with their relative weight is a System of Beliefs.

Probabilities are estimated by the number and weight of the arguments that prove or indicate that a certain thing is, was or will be. Arguments, by themselves, are intrinsic, or artificial in the daily speech, they are expressed or come out according to considerations of cause, the effects, the observer, the connection, indication or any other circumstances that may have any relation with the thing under proof. It can also be external and non-artificial, coming from the observer's authority and his or her witnessing.

The way to apply arguments to conjecture and measure probabilities may follow nine rules or axioms:

1. Applied to things where it is possible to acquire certainty;

2. It is not sufficient to weigh only one or another argument, but it is necessary to investigate all that may arrive at our knowledge and that is appropriate to prove things;

3. One must not consider only arguments that prove something but also those that can lead to an opposite conclusion, so that it becomes clear which one has a bigger weight;

4. In order to judge universalities, remote and universal arguments are enough. However to conjecture about specific things, we must add closer and special arguments if available;

5. In the uncertainty we must cease all our actions until we have more clarity and, if we have to choose between two possibilities we must make the option for the one that may seem more appropriate, safer, wiser or at least more probable, even if none is.

6. What is useful and not prejudicial must be preferred over what has never been useful and is always prejudicial;

7. Human actions must not be evaluated according to the outcome, as, sometimes, imprudent actions have a better result, while reasonable actions may lead to worse results;

8. In our judgments we must be careful before allowing things getting a bigger weight than they deserve, nor to consider something as less probable than something absolutely certain, nor imposing to others the same opinion;

9. Once total exactitude can rarely be achieved we consider as absolute certainty only what is morally correct through need and personalized desire.

The System of Beliefs is thus defined as being the set of arguments and their relative weight. Hence, the amount of information allowed by a set of data about an unknown variable depends on the existent System of Beliefs. Therefore with the same set of data we can obtain differentiated results, influenced by the different Systems of Beliefs we all have and that condition the way we see and how we relate to the world.

Another method, stipulating that before a set of data we can obtain higher information amount is to work and focus in the System of Beliefs… For instance, to question and inquire may be boring because it takes to much work and few results in data obtaining.

To investigate the System of Beliefs, we must raise the following questions (Dorbolo, 2003):

• Which beliefs do we have?

• How do beliefs interrelate?

• How do beliefs relate to our feelings and actions?

- Which beliefs we have is heavier?

- Until how far in past are our beliefs recognizable?

- Where did we achieve our beliefs? Did we create them or did we inherit them?

- Did we experience big changes in our System of Beliefs? How did that happen?

- Is it possible to draw a diagram of a specific cluster of Systems of Beliefs, ideas, feelings or actions?

In short, and the most complete that the set of data is, there will always be a System of Beliefs where the amount of Information over the unknown variable will be zero, due to the limitations of the System of Beliefs in presence of the observer. Many times we are told the way things are but we do not listen.

On the other hand we also have to deal with data uncertainty, as these may be "incomplete, imprecise, fragmented, suspicious, vague, contradictory, or disabled in any other way".

When we approach the matter of information we cannot avoid the observer's role. Carvalho Rodrigues advocates that the measure we take from the amount of information in messages received by our senses is based in perception. He gives us the example of two woks from the Russian painter Kasimir Malevitch[3] (1878-1935) founder of Suprematism (1913), who painted "*A Black Quadrilateral and White over White*", he had enough genius to show in these two works the «the pure feeling supremacy's, or the perception in painting», painting being only a *color construction on a two dimensional space.*

When we ask what is in those paintings the immediate answer is: A black quadrilateral and a white quadrilateral. However we are before a white canvas in the first painting and a canvas with an almost imperceptible line in the second.

It is exactly over these paintings that that we give the black quadrilateral and the white square higher information amount, because they induce in each one of us feelings which in turn generate action. We also know that "*the rarer the event the higher the turmoil in all human structures*".

In this case we were led to conclude that the lesser probability for an event's occurrence, or yet, the lesser its frequency, the bigger our perception upon it, so the higher the amount of information it contains.

"It is this supremacy, the supremacy of our perception over our feelings, the supremacy of information amount we can measure in

an event, over its frequency that commands our behavior that induces major modifications in systems"

(Carvalho Rodrigues, 1989).

---

3 KasimirMalevitch stated that art's reality depends entirely from the effect of color over the scenario. The depicted painting has not a relation of subjection to the real world. It is, in itself and by itself a real fact: it is as "concrete" as all objects surrounding us. It means that the object-painting is not imitating anything: it exists, as objects exist in nature.

It is not surprising that human made creations in the most diversified systems are fragile and precarious and induce great complexity. They are structures influenced too much by non-predictable events, becoming responsible for stronger and everlasting generated perceptions. A system's structure will only be compatible with a proper information amount, otherwise the structure will enter into collapse and that system will have to organize itself with some other structure.

A system with a great amount of data to obtain less information is said to have a highly disordered structure, and induces in the observer a huge ignorance. On the contrary with an orderly and well organized structure, with a set of data with few elements, one can obtain all the necessary information. In short, the degree of knowledge we can have over a determined system is maximized if its uncertainty is zero.

To do so, a method has been developed to calculate the information amount that events generate in the observer, tested in several domains, in studies related with the loss of cohesion from a society before events related to epidemics, or the effects caused in an army before loss in combat, or still with the detection of faults related to the tannery industry, or the determination of fibers distribution in paper industry.

Hence in this chapter's realm we are describing a method to measure information amount. The concept of measure enhances the observer's existence. Hence, the measure of information amount to Simplicity is a relativist measure, it is always dependable on who is effectively observing, because it depends on the observer's System of Beliefs.

In this sense, the choice of what messages or which messages' specter to allow measure may take the designation of those messages as relevant factors to a determined structure within a system.

Hence convictions and concepts into which each observer allows importance, are what we designate as a System of Beliefs. Naturally each observer will have his own, and it is single, and by the time the observer is growing through out his own existence and the world around him, he accumulates ever more beliefs and concepts in respect to serials of things.

One of those things would be, for instance, what the Simplicity's meaning is before a determined individual or group of individuals. Probably meaning can change from one individual to one another. Not in general, but in its specificity, which as we saw can induce to several definitions for Simplicity.

It all depends in the relation on how we perceive and understand the surrounding universe, which takes place in our senses, and naturally in our System of Beliefs. Already mathematician Friederich Bessel, while examining time records on stars transit in Konigsberg observatory, and facing the systematic differences present in observations made by different astronomers, concluded that perhaps he would be before the existence of a "personal equation". Being thus, one can state that an equation is also a personal experience, and it enhances the existence of a sensor that picks up information transmitted by our senses and that our brain processes.

A sensor is something like a device that receives a signal. It's associated to a specific sensation. It receives a signal – *stimuli or data* – and responds through out an electrical signal. Hence understanding that *stimuli or data* are the amount, property or condition detected and con-verted in electrical signal, which in turn is transmitted to the brain by several sensors having the capacity to recognize the surrounding universe.

In short we can state that sensors are responsible by data transmission to the decision maker that in turn consubstantiates into Information.

As we assessed by Maeda's (2006) proposal on the concept of Simplicity, the concept of Information as we understood it is also from a relativist or subjective nature, due to the multiplicity and variety of the existent definitions.

We are thus before a probability: the measure of certainty of who we really are, is obtained by the combination of arguments we presently dispose, knowing that there is a weight for each argument. This set of arguments, plus each argument weight is, in turn, the System of Beliefs (Melo-Pinto, 1998). Finally we can also state that before the same set of data, different Systems of Beliefs will also give different results.

Hence that we can sustain Bernoulli's proposal on arguments combination method, we will now present the results of Melo-Pinto (1998) in the recognition technique (the knowing of what the incognita is) by the mathematical method of combining arguments afterwards developed by Dempster-Shafer (Yager, 1994). They have both defined the set of each argument and its weight as function constituents of the System of Beliefs.

Schools, either from Design or Engineering, were created as to allow the same school's System of Beliefs to deal with design or the design with the fewer possible data.

The question now raised is how in Design or Engineering we built a System of Beliefs to allow design or the design simpler in itself and for users? We can state that it can be made through combined arguments or weights that are part of the System of Beliefs.

5. Decision processes: Combining arguments

According to Melo-Pinto (1998), decision process results from information gathering driven to that decision. But all along, the natural appearance of new arguments (partial or not) may drove us to remake beliefs by the light of new data.

Still according to Melo-Pinto (1998), who developed a system of decision making applied to the visual recognition of images in degrading situations, he states that *due to different beliefs' functions over the same insight system, based on different bodies of evidence, he supports himself in the Dempster-Shafer rule of combination*, and as such, may help to calculate the respective functions of believe and plausibility. However, as that function is probabilistic it should be evidence resultant. But Melo-Pinto contradicts this fact, stating in turn that value associated to a given image effectively renders the beliefs we have of it, and are independent from evidence. He gives, for instance, the case of a blond woman's image (Marlene Dietrich) from whom we need no additional result to sustain how associated that image is to Marylin Monroe. Hence the capacity of combination is fundamental to a method that deals with different argument.

"Design Universal Principles" may be those "arguments" than can be embodied as Design essence. They are, no less, and according to Lidwell et al. (2010), the meanings we give to the artifacts we use, in usability terms as well as in its influence, perception and usage call in everyday life, and can assume their selves as combined data and arguments.

As seen before, different arguments combining with a specific weight is what embodies our system of beliefs.

The issue we could now raise in this chapter is what is the reason why some products are most desired by some certain people over others? Which method could be used as to allow knowing which arguments each one of us gives more or less relevance?

Today we know that the supposed commercial failure of some products is due not only to problems related to an operative-functional nature, but also because they didn't made sense to the aimed target. Hence we can state that this is a consequence of something that does not fit with the positivist perception of a certain user.

In that sense, this chapter now aims to let know the "Design Universal Principles" (Lidwell et al., 2010), as a first guide to interdisciplinary reference not only to designers but also to engineers as it combines a vast set of arguments in the shape of 125 concepts related to Design and also to Psychology, Engineering and Architecture, organized through five categories, it can be used as a guide to structure the conception for Simplicity, not only for Design, but also as a reference for Engineering:

• How can Design and Engineering's perception be influenced?

• How to help people learn about Design and Engineering?

• How to improve Design and Engineering's usability?

• How to raise the call for Design and Engineering?

• How to improve decision taking in Design and Engineering processes?

These five categories are in turn subdivided into 225 contents that raise questions as diversified as: accessibility, archetypes, linings, cognitive dissonance, color, comparison, confirmation, consistence, convergence, cost-benefice, development cycle, errors, safety factor, Fibonacci's sequence, figure-background relation, usability-flexibility, forgetting, form and function, Gutenberg's diagram, hierarchy, highlight, iconic representation, interference effects, Prägnanz' law, legibility, life cycle, mental map, modularity, normal distribution, among many other, and truly it reflects a decision process, as it has the capacity to combine some of the mentioned arguments that may simplify processes while elaborating a Design or Engineering design.

But is it possible, as we call upon those 225 principals, that we became paralyzed while design acting? After all, where to start designing, which or what aspect should we allow more or less importance? Supposedly hardly anybody will be able give the answer. Even so, believing the existence of those "beliefs", there can only be, at the most, one or two... Hence, assuring the 225 (15^2) "principals" will not be in fact a statement to consider.

What will make sense is to state, that at the most, there was someone who described 225 arguments or weights to build or describe his own system of beliefs.

Hence the concept of Simplicity in Design or Engineering is the one in which effectively each observer beliefs as its own, and that it naturally depends on the amount of information he has relatively to the data he possesses, as the result of combining different arguments or data:

Set of data C

$$C = \{C_i\} \qquad i = 1, m \tag{3}$$

Set of data D

$$D = \{D_j\} \qquad j = 1, n \tag{4}$$

Maximum Information Amount given by set of data C

$$H_i = \sum -p_i \log p_i \qquad i = 1, m \tag{5}$$

$$H(C) = \sum H_i \qquad i = 1, m \tag{6}$$

Maximum Information Amount given by set of data D

$$H(D) = \sum H_j \qquad j = 1, n \tag{7}$$

$$H_j = \sum -p_j \log p_j \qquad j = 1, n \tag{8}$$

$p_i \wedge p_j$ are the sample case $C_i \wedge D_i$ respectively.

Before a system of beliefs Bel_i in which $i = 1, m$ the information amount allowed by the set of data C is:

$$H_{ii}(C) = \sum Bel_i \otimes H_i \qquad i = 1, m \tag{9}$$

Before a system of beliefs Bel_k in which $i \wedge k = 1, m$ the information amount allowed by the set of data C is:

$H_{ki}(C) = \sum Bel_k \otimes H_i,$

$$k = 1, m \wedge i = 1, m \qquad (10)$$

If $H_{ii}(C) < H_{ki}(C)$, then the system of beliefs Bel_k is better than the system of beliefs Bel_i for the set of data C.

If $H_{ii}(C) > H_{ki}(C)$ is the system of beliefs Bel_i is better.

If $H_{ii}(C) = H_{ki}(C)$ are Bel_i and Bel_k equivalent to the incognita knowledge.

The same is applied to the set of data D.

$$H_{jj}(D) = \sum Bel_j \otimes H_j, j = 1, n \qquad (11)$$

$$H_{kj}(D) = \sum Bel_k \otimes H_j \quad k \wedge j = 1, n \qquad (12)$$

If $m < n$ and $Bel_i = Bel_k$ when

$$H(C) \geq H(D) \qquad (13)$$

It indicates that for the system of beliefs Bel_i C we need less number of elements than the set D to give higher information amount about the unknown variable. Hence, C is simpler.

If $m < n$ and $Bel_i \neq Bel_k$ when

$H_{ii}(C) \geq H_{jj}(D) \Rightarrow C$ is simpler.

If $m < n$ and $Bel_i \neq Bel_k$ when

$$H_{ii}(C) \leq H_{jj}(D) \qquad (14)$$

Bel_k system of beliefs makes set D the simpler set though with more elements than set C.

If $m < n$ and $Bel_i \neq Bel_k$ when

$$H_{ii}(C) \geq H_{jj}(D) \qquad (15)$$

Bel_i system of beliefs extracts more information amount than set C with fewer elements than D. C is simpler.

If $m < n$ e $Bel_i \neq Bel_k$ when

$$H_{ki}(C) \geq H_{kj}(D) \tag{16}$$

C is to system of beliefs K the simpler. It has less elements and allows higher information amount about the unknown.

If $m > n$ and $Bel_i \neq Bel_k$ when

$$H_{ki}(C) \leq H_{kj}(D) \tag{17}$$

Bel_k system of beliefs makes set D the one that gives higher information amount about what we do not know.

In conclusion we can state that for the same system of beliefs the set that gives higher information amount is the simpler.

For the same set of data the system of beliefs that allows a set of higher information amount about the incognita is the most adequate to predict.

One can thus deduct that simpler is what needs less number of elements or set of data to obtain the same information amount or even higher.

This is the essence of Simplicity, and it can be sustained in "Design Universal Principles", in the quality of an interdisciplinary reference guide, either in Design as in Engineering.

However and as complete as it may be, the set of data that can be involved in the conception of a project of Design or Engineering there will also and always be present a System of Beliefs built over a given observer, to whom the information amount over the unknown variable will be zero, as also, there will be a system of beliefs that before a very incomplete set of data will obtain the maximum information amount over the unknown variable.

We can thus deduce that Design in particular and its several schools are liable to be related to mathematics. That schools are a system of beliefs, and they must be not simply a place for knowledge transmission but also a place that promotes knowledge emergence associated to Simplicity.

We will be, as much, before arguments or weights that each School created to give sense to its own design, in this case having as basis the combining arguments according to Dempster-Shafer theory.

Truly a designer can assign determined values to the arguments or weights in cause and another designer can assign to the same arguments or weights another set of values.

Hence Design is no longer a derivation from art, not even from object engineering (MOURA, 2012). Design can be what informs about human creativity, that emerges from the combination and interaction between the several fields of creativity itself, to what we would like to add that it depends on the combining of different arguments and weights of the system of beliefs to what that designer pertains and that will always be the one he received in its School as a student.

We here designate as pertaining to a same school all those who share the same arguments with the same weight.

Hence we can state that we can have as much Schools or Systems of Beliefs as the nature of arguments (Ma) or weights (P):

$$\{Ma_i, P_i\} = E_i \tag{18}$$

$$\{Ma_j, P_j\} = E_j \tag{19}$$

$$\{Ma_k, P_k\} = E_k \tag{20}$$

So that a School can be shaped:

$$Bel_{i,k} \bigcup Bel_{i,l} \bigcup Bel_{i,m} \Rightarrow Bel_{i,k} = Bel_{i,l} = Bel_{i,m} \tag{21}$$

when:

$$\frac{m}{n} \cong 1 \tag{22}$$

The result is that a System of Beliefs is a School, a cluster, and it shares the same arguments with the same weight allowed to the same arguments:

$$Bel_{i,k} = E_i \quad k = 1,...,m \tag{23}$$

Hence we can deduce that Schools have the gift to simplify things. And that the set of data or arguments is equal to 1. Naturally the better the School the less data it will need to explain or transmit its knowledge.

6. Conclusion

In conclusion, some aspects can be evidenced:

a. The essence of the meaning of "Simplicity" is not one of easy understanding in the way it has been, so far, described. The common knowledge is that there are undetermined number of definitions to explain what effectively is Simplicity;

b. That "Simplicity is relativist", as an example of "easiness", of "difficultness", of "information", or "complexity", among others, are issues of each one measurements. As a system of beliefs it will always be associated with the observer.

c. That like "Simplicity", until now the concept or notion of "Information" is vague and intuitive.

d. That before a huge Simplicity, we will be before a major amount of information, with restricted data;

e. That Design in particular and its several schools are also passable of "mathematization";

f. That Maeda's laws to describe "Simplicity" are a set of arguments and weights built in Maeda's System of Beliefs;

g. At last that a school is always committed to its shared System of Beliefs, and as such it is responsible for values, ideals, feelings and actions transmitted to those who are or were, part of itself.

In conclusion we can also state that Maeda's laws are not effectively laws, but methods or suggestions to define simplicity, because as we saw the simplicity measure explained does not need further increase.

Hence, if we use this relativist measure for simplicity coupled with the complexity measure, we then have the quantitative parameters that release us of qualitative laws for engineering or for design. They allow us the necessary amount to assess design, or engineering, or both.

Therefore the encounter between complexity and simplicity is an art, once we are capable of measuring it and, as such, to quantify it, we acquired the necessary tools for either an engineer or a designer, to achieve the objective of conceiving and build concepts, methods, processes or products, either organic or functional, and hence, at the very end this is what an engineer or a designer aims to. Complexity and simplicity is an art over which we are capable to put measures and, as such, to quantify.

In conclusion we dare state that if we were in the presence of a philosopher; he would say that the true definition for simplicity would be: it exists as to make sense within the world that surrounds us.

It will be then demonstrated that "Simplicity" (its measurement) in particular that of Design, being measurable, is also an Engineering parameter.

Acknowledgements

Thanks to:

Professor F. Carvalho Rodrigues was responsible for encouraging and helped me the fulfillment of this vision.

Martim Lapa, for his admirable staying power to support me to translate this version.

Professor Denis A. Coelho for editing the manuscript.

Finally, I would also like to thank the UNIDCOM/IADE.

Author details

Carlos A.M. Duarte*

Address all correspondence to: carlos.duarte@iade.pt

IADE – Creative University, UNIDCOM - The Research Unit, Lisbon, Portugal

References

[1] BernardClaude (1957). *An Introduction to the Study of Experimental Medicine*. Dover Books on Biology [Paperback].

[2] BernoulliJacob; *trad.* SYLLA, Edith Dudley (2006). *The Art of Conjecturing together with Letter to a Friend on Sets in Court Tennis*. The Johns Hopkins University Press, Balti-more, Maryland.

[3] BernoulliJakob (2010). *Ars Conjectandi* (The Art of Conjecturing). [s.n.].

[4] Carvalho RodriguesF. (1989). *A Proposed Entropy Measure For Assessing Combat Degra-dation*, J. Opl. Res. Soc. 40, (8).

[5] CreswellJohn W. (2009). *Research Design: qualitative, quantitative, and mixed methods ap-proaches*. SAGE Publications, Inc.

[6] Fisher, R. A. *ed.* Bennett, J. H. (1990). *in Oxford Science Publications* "Statistical Meth-ods Experimental Design and Scientific Inference". Oxford University Press.

[7] GanorBoaz; von Knop, Katharina; Duarte, Carlos (2007). *Hypermedia Seduction for Ter-rorist Recruiting*. IOS Press.

[8] GeneMike (2007). *The Design Matrix, a consilience of clues*. Arbor Vital Press.

[9] LidwellWilliam; Holden, Kritina; Butler, Jill (2003). *Universal Principles of Design: 125 ways to enhance usability, influence perception, increase appeal, make better design decisions, and teach through design.* Rockport Publishers.

[10] LindsayPeter H., Norman, Donald A. (1977). Human information processing: An introduction to psychology, Academic Press.

[11] MaedaJohn (2006). *The Laws of Simplicity.* Massachusetts Institute of Technology.

[12] Melo-PintoPedro (1998). *Aplicação da Teoria das Crenças ao Reconhecimento Visual,* Universidade de Trás-os-Montes e Alto Douro, Vila Real.

[13] NormanDonald A. (2004). *Emotional Design.* Basic Books.

[14] Shafer, G. (1978). Non-additive probabilities in the work of Bernoulli and Lambert, *Archive for History of Exact Sciences,* 19 , 309-370.

[15] Shannon, C. E. (1948). A Mathematical Theory of Communication". The Bell System Technical Journal, 27, July- October., 379-423.

[16] YagerRonald R.; Fedrizzi, Mario; Kacprzyk, Janusz (1994). *Advances in the Dempster-Shafer Theory of Evidence,* Wiley.

Sustainability

Chapter 7

Toy Design Methods:
A Sustainability Perspective

Denis A. Coelho and Sónia A. Fernandes

Additional information is available at the end of the chapter

1. Introduction

The challenges of sustainability are already guiding today's scientific and technological progress, and because environmental awareness should necessarily be fostered from birth, toy design was selected as a field of inquiry for this contribution. The connection between toys and environmental sustainability is explored in this chapter in two main areas which complement each other. On the one hand, eco-design of toys is considered "tout court" (pure and simple). In this respect, the ecological aspects of the objects that support play (toys) should be considered at the level of the materials used, the expenditure of energy in their production and transportation, as well as in what concerns other issues related to the lifecycle of the product and for a combination of recreational, educational, and pedagogical purposes. On the other hand, children may also be actively educated to develop their environmental awareness. It is believed that these two approaches are complementary and that in an ideal scenario the toys for leisure activities given to children by their relatives, and those provided by nurseries, day-care centres, pre-schools, primary schools and even urban equipment available for kids in playgrounds should reconcile these two aims: minimizing environmental impacts (associated with pro-active eco-design, considering the whole life cycle of the recreational-educational product) and education for civic development in view of responsible citizenship with great emphasis on the foundation of environmental awareness and protection of ecosystems and the legacy for future generations.

To contextualise and underpin the development of this work, it proceeds with an initial gathering of information about the child's cognitive development, which is presented in the reminder of this section. The following sections seek to answer the questions raised as a result of the proposed aims along with the presentation of design results.

2. Stages of child development

A child is a human being in the "cradle" of her or his development. Childhood is the period from birth until about the 12th year of a person's life. It is a time of great physical development, manifested by progressive increase in height and weight of the child. It is also the period when the human being develops psychologically, and during which changes occur in behaviour and the foundations of personality are developed. Regarding the maturity of all species that inhabit the Earth, the human being is the one with the slowest growth and development rates of all. The human is a rather slow developing species; for example, in the time that a child learns to walk and run with sufficient equilibrium, other species reach full maturity, as is the case of rats, who reach sexual maturity in only 15 days.

According to Wallon (1981), the human being is determined physiologically and socially, subject to internal arrangements (affective), and external situations (sensorial-motor). In this way the study of human development must consider the subject as grounded in its relationship with the environment. This author considers the following five stages of development: impulsive-emotional - 0 months-1 year (the predominant affection); sensor-motor and projective - 3 months-3 years (dominated by intelligence); personalism - 3-6 years (formation of the personality of the individual and self-awareness); categorical stage - 6-11 years (development of memory capacities and voluntary attention); stage of adolescence - 11-16 years (physical and psychological changes). However, the stages of human development do not cease in adolescence, as, according to Wallon (1981), the learning processes that occur throughout life involve crossing a new stage of development.

Jean Piaget considered that the development of children also occurred in stages, however, according to Piaget and Inhelder (1995), the key is the sequential order of stages and not the age at which each one arises. To make up a new stadium, one must have passed through and overcome the previous stages. Jean Piaget believed that there are four stages of development, which he detailed in his Cognitive Theory. These stages are: the sensor-motor stage - 0 months-2 years; preoperative - 2-7 years (egocentricity); stage of concrete operations - 7-11 years (integrated mental organization); and stage of formal operations - 12 years and over (development of abstract thinking operations).

2.1. Observation of children at play

The previous section, aimed at contextualizing the work in this chapter, which is centred on children, presented the stages of their cognitive development (psychosocial and emotional). In order to support this contextualization, a set of observations of children at play, presented in Table 1, were made, as a form of recognition in practice of the concepts presented. These observations also allowed inference of the type of activities, the duration thereof and the reactions of children during play. Especially in the cases where two children of different ages were observed playing together, the observed states of the children are different in some cases. These result from the diverse stages of development of the children and the adequateness and appeal and complexity of the activities being carried out.

date	time	actors (age)	Description of activities	General impressions per child	
24-12 2010	17:12	Leandro (23 m.) Leonor (4 y.)	- hide and seek (33min) - sketching and colouring (1h14min) - clay moulding of dolls (25min)	**Leandro** - interest or joy - "difficulty" -indifference	**Leonor** - energy or joy - concentration - tiredness
26-12 2010	14:46	Leandro (23 m.)	- ball play (24min) - tricycle riding (14 min) - looking at pets (caged birds) (12min)	**Leandro** - joy or motivation - joy or motivation - apprehensive or fearful	
28-12 2010	15:30	Vasco (14 m.)	- looking at books with sounds and colours (12min) - puzzles and cubes (40min) - playing with the dog (15min)	**Vasco** - concentration or joy - interest or concentration - joy or satisfaction or happiness	
02-01 2011	10:53	Catarina (3 y.)	- looking at books (38min) - using paint and scribble books (45min)	**Catarina** - interest - happiness	
15-01 2011	14:25	Tiago (12 y.) Alexandra (10 y.)	- playing playstation (1h) - ball games (35min) - pretending game (20min) - reading (30min) - playing with dolls and cars (15min)	**Alexandra** - indifference - joy - "imagination" - interest - "imagination"	**Tiago** - concentration - energy or satisfaction - "imagination" - indifference - "imagination"

Table 1. Observation of children at play (several children interacted with the second author in a family setting).

3. Education for sustainability

In order to find a point of connection between sustainability and education for sustainability a search was initially made for concepts and needs within sustainable development. Subsequently, an analysis was made of the universe in education for sustainability and a letter from the Earth is presented, which is a declaration of fundamental principles for building a society that is fair, sustainable and peaceful.

3.1. Sustainable development

Enough for everyone, forever. These words resonate with the ideas of limited resources, responsible consumption, equality and equity and a long-term perspective, all of them corresponding to important concepts of sustainable development (Portuguese Ministry of

Education, 2006). Sustainable development involves meeting present needs without com-promising the ability of future generations to also meet theirs. The concept of sustainable development involves not only fostering positive impacts locally, but this should be thought in global terms, giving rise to the sense of universal responsibility. Hence, sustainable devel-opment is all about a joint effort carried out between different areas, whether social, eco-nomic, ecological or political, thus trying to strike a balance between economic growth, social equality and the preservation of natural resources and habitats. Ensuring that the peo-ple from all over the world are able to satisfy their basic needs, while assuring that future generations can have the same quality of life is at the core of the sustainable development agenda.

3.2. Education for sustainable development

'The early childhood years are the most significant and when the greatest developments in the life of a person take place and are generally regarded as the foundation upon which the rest of an individual's life is built (Mustard 2000; Rutter, 2002). Children, viewed beyond their genetic heritage, are influenced by the environment around them and by their relation-ships with their parents and with other people, so when considering the formation of emo-tions in children, one must look at all the ways in which the child responds to all persons with whom he or she crosses, and all the images she or he sees. The education of children is the greatest responsibility of parenting, with the aim to create "adulthood" and not perpetu-ate childhood, parents should bet early on in showing their children the emotional realities with which life confronts everyone, sooner or later.

As is well known, children follow the examples of parents and of all those who are part of their reality. It is however up to parents to ensure the safety of small children that are not yet able to assess the dangers that surround them, and to educate them to become citizens of a just society where everyone can exercise their rights to equality and solidarity. One ought to never forget that home is the real trainer of people. Values such as education and person-al development must be transmitted within the family, however, in current times, and with the economic and social situation most people are confronted with today, it becomes in-creasingly difficult. Parents spend less time with their children, thus hindering the teaching of personal and social values which includes education for sustainable development. To compensate this, the school seeks to fill the gaps in education by the family, but there is still a need for restructuring at the curriculum level. The school curriculum structure does not necessarily facilitate the task of educating for sustainable development. For example, experi-mental teaching activities can be a real challenge when teaching is confined to a classroom. Furthermore, the assessment systems are, often, based in a competitive model in which indi-vidual grades become the main goal of the students. This is, in fact, an environment that hin-ders the promotion of values central to the notion of sustainable development, such as 'participation' or 'cooperation' (Portuguese Ministry of Education, 2006).

The call for a rampant consumerism that translates into large discharges of toxic waste and garbage are two major global problems. Through early education for sustainable development it may be possible to modify consumer relations and ensure environmental sustainability of our planet changing the current situation.

Section	Principles
I - Respect and care for community life	Respect Earth and life in all its diversity. Care for the community of life with understanding, compassion and love. Build democratic societies that are just, participatory, sustainable and peaceful. Conserving Earth's bounty and beauty for present and future generations.
II - Ecological Integrity	Protect and restore the integrity of Earth's ecological systems, with special attention given to biological diversity and natural processes that sustain life. Addressing the prevention of environmental problems as the best method of environmental protection and in case of insufficient recognition, taking preventive measures. Adopt patterns of production, consumption and reproduction that safeguard Earth's regenerative capacity, human rights and the welfare of communities. Encourage the study of ecological sustainability and promote the free exchange of knowledge and its application.
III - Social and economic justice	Eradicate poverty as an ethical, social and environmental problem. Ensure that economic institutions at all levels promote human development in an equitable and sustainable manner. Affirm gender equality and equity as prerequisites to sustainable development and ensure universal access to education, health care and employment. Defend, without discrimination, the right of everyone to a social and natural environment, by promoting human dignity, bodily health and spiritual well-being, with special attention given to the rights of indigenous peoples and minorities.
IV - Democracy, nonviolence and peace	Strengthen democratic institutions at all levels, and provide transparency and effective governance to ensure inclusive participation in decision making and access to justice. Integrate knowledge, values and skills for a sustainable way of life into formal education and lifelong learning, Treat all living beings with respect and consideration. Promote a culture of tolerance, non-violence and peace.

Table 2. Earth Charter - basic principles (Portuguese Ministry of Education, 2006).

3.3. Analysis of the educational universe in the area of sustainability

In Portugal, there are a set of pedagogical guidelines to support education for citizenship, which are published in the booklet entitled "Guidelines for Education for Sustainability", developed jointly by the Directorate General for Innovation and Curriculum Development, Ministry of Education and the Portuguese Association for Environmental Education (AS-PEA). Its realization is based on the Earth Charter (Table 2), which was published by UNESCO in 2000 and approved by the UN in 2002. Published in 2006, this guide aims to foster and support primary school teachers in the arduous task of educating for sustainable development, and in serving as a basis for curriculum and civic education of children and youth. "The authors believe that the school, among other institutional actors, plays an invaluable training role that must be exercised and enjoyed in large areas, including not only formal knowledge and curriculum (...) "(Evaristo, 2006). Making the school a hub for production and dissemination of information on education for sustainable development for students and parents is one of the objectives of ASPEA.

The Earth Charter is a reference to relevant and unique training programs that aim to develop learning processes in students for a more just, sustainable and peaceful society, (Portuguese Ministry of Education, 2006).

The literature review of concepts related to education for sustainability and sustainable development, contributed significantly to the design phase of this work. In Portugal there was already a breakthrough in early education for sustainable development, which seeks to instil values such as early environmental sustainability in school children. However, authors can see that this whole journey tends to occur mostly at the theoretical level, neglecting the practical part as a crucial incentive. It is worth highlighting the commitment and positive attitude of the Ministry of Education to prepare a script that aims to guide teachers in the arduous task that is education for sustainable development.

4. Methodology for toy design

The literature review carried out to meet the goal set for this work, of reviewing existing methodologies for the design of toys, was not successful, since it fell through without attaining literature references covering this matter. The authors proceeded to propose a methodology for designing new toys (based on the systematic process of design and taking into account the stages of child development) that is the following:

1. Recall the stages of physical, cognitive, sensor-motor, social and emotional development of children.

2. Given the context of playful activities, proceed to carry out an exploration of activities that may contribute to the development of the child in one or more of the spheres covered above (1).

3. Find one or more metaphors that may form the basis of concepts for the creation of toys or recreational objects.

4. Evaluate the concepts of toy or object triggered as a result of the previous stage in satisfying a set of requirements generally applicable to toys or objects of play (e.g. low toxicity, safety regarding self-inflicted injury) and select those that satisfy the general requirements and that are configured as original proposals, potentially motivating their use by children (by selecting different age groups) and clearly support one or more activities that promote psychosocial development, development of sensory-motor skills and of physical ability in the child users.

5. Develop and set, based on knowledge of the context of child development and the concept selected, a specification in order to guide the design of the toy or playful object. At this stage market objectives should be considered, including costs, packaging, distribution and consideration may also be given to objectives of another nature.

6. Proceed to the development of the detailed concept and produce prototypes enabling testing under controlled conditions of safety, initially with adults and ensuring no hazard is presented by the prototypes when seeking to involve children in their use. (Note: at this stage children should be able to keep the prototype toys).

7. In this process the results of usability testing can motivate changes to the project description and a new iteration of design refinement and testing, reiterating until the development team is satisfied with the results, or the resources allocated to the development have been exhausted.

8. Development of production processes and of release, distribution, and marketing documentation.

4.1. Evaluation of the proposed methodology

Since there was no published methodology found alluding to the subject discussed in this chapter with respect to the design of toys, authors opted for the generation of a new methodology. The assessment of this proposed methodology was achieved through the implementation of projects based on it. Table 3 describes the strengths and weaknesses found in the pursuit of the conceptual design phase, which includes the first four steps (Table 3).

Given the academic nature of the development of plans to evaluate the proposed methodology, steps 5, 6, 7 and 8 could not be tested fully. However, step 6 was partially implemented, given the production of two prototypes, without adopting the colours, materials and the final dimensions of the toys that were designed.

Not having found a published methodology, the authors chose to develop a methodology for designing toys focused on the stages of psychosocial and emotional development of children from an early age. One of the most prevalent weaknesses of the methodology relates to the initial survey to be carried out about the stages of physical, cognitive, sensor-motor, social and emotional development of children, which may be considered as limiting creativity. As strengths, the proposed methodology's capacity to foster iteration and improvement after the prototyping phase is highlighted.

4.2. Toy design methods considering sustainable design goals

The sustainable design methodologies proposed by Fuad-Luke (2004) and Ryan (2009) were taken as a basis on which to develop a proposal for toy design satisfying sustainability goals. The former is rather more detailed than the latter, with a high level of detail given to the process, which is deemed easy to follow. As a strong point in Ryan's (2009) proposal, authors emphasize the fact that not only does it cover a perspective focused on the product but it also encompasses product and service systems with strategic orientation towards sustainable design goals. Both methodologies share the common goal to create products or artefacts that safeguard the continuity of the planet's resources, thereby creating a combined economic, social, and environmental solution. The concern with the product life cycle is also a common point in both methods. Fuad-Luke (2004) presents a methodology for the eco-pluralist designer, easy to understand, so that designers can design more sustainable products aimed at the continuity of future generations. Ryan (2009) proposes, on the other hand, a more elaborate method dealing with systems and that, as such, can be adopted and used by companies. Considering the initially proposed methodology for toy design (Table 2) and the contributions reaped from Fuad-Lake (2004) and Ryan's (2009) methods, the toy design methodology proposed was enlarged towards being geared towards sustainability goals and towards fostering the development of environmental awareness, taking the form presented in Table 4.

Step of the proposed methodology	Strengths and Weaknesses
Step 1	Strengths: With the implementation of this step, one gets a very comprehensive view of the stages of child development. Weaknesses: Once you start the development of this methodology with a focus on literature, from the standpoint of design activity, creativity is a bit on standby as it is not part of the realization of sketches. Instead of only collecting data from literature, it might be more stimulating and creative to simultaneously consider the relevance of designed objects to interact with children.
Step 2	Strengths: Given the focus of activities that contribute to the development of specific capabilities of the child, the methodology can be used many times by various designers giving rise to very diverse projects. Weaknesses: Since the methodology focuses on specific activities related directly to child development, the results may be relevant only to a very narrow age span, challenging the duration of the interest of the child in the toy over an extended period, which may undermine the objectives of sustainability.
Step 3	Strengths: The use of metaphors opens up almost unlimited possibilities. The simultaneous use of more than one metaphor is intended to prevent that a metaphor may predominate and the subject may become too literal if using only one metaphor. The crossing and the combination of several metaphors is a way of stimulating creativity, enhancing innovativeness of results. Weaknesses: If the designer is not careful, the project may become too literal in relation to the metaphor, so the designer must be aware and avoid over-literalness.
Step 4	Considering the initial proposed methodology of the concept generated is a way to avoid that the project goes much forward before judging its relevance, which contributes to increase the efficacy of the methodology and to reduce costs (for example in prototyping), and time spent by the designer, or by the design team, in creating the toy. Weaknesses: The focus on specific activities and sensory-motor skills as well as on differentiation into age groups may not be possible given that the concepts generated have been from the outset (in Step 2) directed to a specific activity focusing on one age group and supporting the development thereof.

Table 3. The authors' evaluation of the proposed methodology for toy design.

Step	Activities
1	Review the stages of physical, cognitive, sensor-motor, social and emotional development of children to meet real needs rather than needs related to passing fashion or driven by the markets.
2	Given the context of the design project being developed, carry out exploration activities that may contribute to the development of the child in one or more of the spheres covered above (point 1).
3	Generate concepts for one or more activities that can underpin the creation of toys or recreational objects, directing creativity to issues that foster awareness of environmental sustainability in an educational manner: a - Do not waste materials, energy, food ... b - Respect ecosystems. c - Preserving the planet for future generations. d - Adopt the idea of the three R's - Reduce, Recycle, Reuse. e - Strengthen the relationship between economics, technology, society, politics and the environment. f - Enter the challenge of "moving from concept to action". g - Recover and develop values and behaviors such as trust, mutual respect, responsibility, commitment, solidarity and initiative.
4	Evaluate the toy or playful object concepts triggered in the previous stage against a set of requirements generally applicable to toys or playful objects (e.g. low toxicity, safety against injury) and select those that satisfy the general requirements and are configured as original proposals, potentially motivating their use by children (selected according to different age groups), and that satisfy in an obvious manner the support of one or more activities that promote the development of psychosocial and sensory-motor skills as well as physical ability.
5	Set, based on knowledge of the context of child development, a specification of the concept selected in order to guide the design process of the toy or playful object. At this stage market objectives should be considered including cost, packaging, distribution and consideration may also be given to objectives of a different nature, including sustainability, and such as: a - Designing to minimize the ecological footprint of the product, material or service, that is, reduce the consumption of resources including water and energy. b - Designing to take advantage of renewable energies (solar, wind, hydro or wave), instead of using non-renewable natural capital such as fossil fuels. c - Designing to enable separation of the components of the product, material or service at the end of their life-cycle in order to encourage recycling or reuse of materials and, or, of the components. d - Designing to eliminate the use of toxic or hazardous substances for humans and other life forms in all stages of the life cycle of the product, material or service. e - Designing to engender maximum benefits to the intended audience and to educate the client and the user and thereby create a more equitable future. f - Designing to use locally available materials and resources whenever possible (think globally but act locally). g - Designing modularly to encourage and allow sequential purchases, as required and according to financial availability, in order to facilitate repair and reuse and improve functionality. h - Designing to create more sustainable products, materials and services for a more sustainable future.
6	Review the existing product market, including environmental and social features.
7	Developing a picture of the profile outlined by the environmental impact of the new product.

Step	Activities
8	Proceed to the development of the detailed concept and produce prototypes enabling testing under controlled conditions of safety, initially with adults and ensuring no hazard is presented by the prototypes when seeking to involve children in their use. (Note: at this stage children should be able to keep the prototype toys).
9	In this process the results of usability testing can motivate changes to the project description and a new iteration of design refinement and testing, reiterating until the development team is satisfied with the results or the resources allocated to the development have been exhausted.
10	Development of production processes and of release, distribution, and marketing documentation.

Table 4. Methodology for the design of toys that promote awareness of environmental sustainability.

Starting from a methodology proposed for toy design with 8 steps (Table 2), a new methodology in 10 stages was proposed in Table 4, by agglutination of steps aimed at reducing environmental impacts and at promoting awareness of environmental sustainability. Environmental considerations were introduced in step 5 (which is new), in the sixth step (which was previously step 5) and in step 7 (new). The methodology for toy design presented is aimed at promoting awareness of environmental sustainability. This is based on a systematic design process, including steps to lead a process of sustainable design of toys with inclusion of actions meant to integrate themes closely linked to education for environmental sustainability.

4.3. Toy design - 1

In applying the new methodology for the design of ecological toys, a concept was generated and implicitly chosen that underwent several iterations in order to adapt it and make it compatible with the safety requirements associated to these kinds of objects. The authors opted for the choice of children aged from 1 to 2 years, which is regarded as a phase of great sensory and cognitive development of the infant. The activity on which to focus the project on which the authors decided to implement the proposed methodology was the activity of fitting between parts, for which a preferred order is defined which will be a secondary challenge (the primary challenge is the realization of the fitting). In the project developed, authors also considered the learning activities leading to colours identification.

The authors considered several metaphors in the development of the project, having been incorporated in the foreseen interaction of the user with the toy, the fitting of cups and throwing rings into a pole, as well as a fruit tree.

The initial concept renders are presented in Figures 1 and 2. After the smoothening of sharp edges was done, and after creating empty spaces to make the parts lighter and less bulky, the same concept evolved and gained the aspect that can be appreciated in Figures 3 to 5 (the smoothening of edges aimed at satisfaction of the requirements associated with safety against damage to kids made by themselves).

Figure 1. Initial concept renders (a) – toy design – 1.

Figure 2. Initial concept renders (b) – toy design – 1.

Figure 3. Evolved concept (a) – toy design – 1.

Figure 4. Evolved concept (b) – toy design – 1.

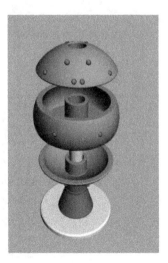

Figure 5. Evolved concept (c) – toy design – 1.

For the actual manufacture of the toy it was decided to select a biodegradable material such as natural latex. This is a flexible material, it does not hurt upon hitting it suddenly and there is no risk of falling as toys are spread into the room. Its production is carried out by

moulding and casting, followed by curing in oven drying. For the colouring, non-toxic wa-
ter-based pigments were chosen.

The prototype of the first design outcome of the project was produced by a three dimensional printing process (based on gypsum) and was made on a scale consistent with the capabilities and limitations of the 3D printer available to the authors. Figures 6 to 8 are images of the
results of prototyping.

Figure 6. Model (a) of toy design – 1.

Figure 7. Model (b) of toy design – 1.

Figure 8. Model (c) of toy design – 1.

The first project aims to be an example of green design features due to its biodegradable material. It is also looking into cultivating small children's early sensibility to care for Nature and for our planet. Despite their young age, most child users are already able to recall small actions and replay them later.

4.4. Toy design - 2

The second project, which is based on methodology that aims at education for awareness of sustainability, incorporates the principles of respect for ecosystems and development of behaviour and values. This toy consists of a small tree that is inhabited by four different animals, and seeks to show that everyone is entitled to their space. Actions such as deforestation often cause the extinction of animal species and natural habitats, and as such, it is intended with this toy that the little kids become interested in the continuity of the planet.

According to the approach presented in Table 4, the authors proceeded with the development of project activities aiming to design a toy for children seeking the development of environmental awareness. The stage of cognitive development targeted was from 3 to 5 years and the activity triggered was the development of recommended levels of membership and association through stimulating of the recognition of the compatibility of symbiosis, including engagement between peers. For the generation of concepts, respecting ecosystems and recovering and developing values and behaviours such as mutual respect and commitment was specially taken into account. Over several drafts and sketches the authors explored various ideas for this project. The concept that came to be developed was based on the idea selected from among many ideas generated. The prototype was performed in 3D printing of high quality ceramic material. In the following images, the prototype is represented as image renders of the 3D model (Figures 9 and 10) and photographs of the prototype (Figures 11 and 12).

Figure 9. Render of concept (a) – toy design – 2.

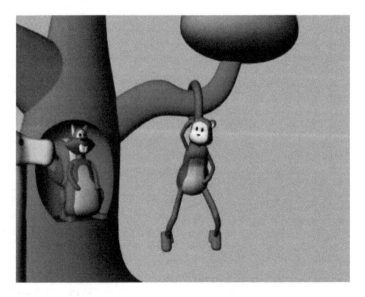

Figure 10. Render of concept (b) – toy design – 2.

Figure 11. Model (a) of toy design – 2.

Figure 12. Model (b) of toy design – 2.

5. Conclusion and future work

In this chapter, design as a driver for education for environmental sustainability and the need for application of ecological concepts in the production of new products, in this case in what concerns toys, have resulted in the creation of new toy design methodologies.

The methodologies presented in this chapter were the result of a lengthy and elaborate research about early childhood education, environmental sustainability and sustainable development and of research on products already produced and which are available.

The methodologies for the design of toys and ecological design for education awareness for environmental sustainability are the result of an amalgamation of three methodologies, thus trying to bridge existing gaps in literature.

The chapter concludes with presentation of two toy projects that are expected to contribute to the aims set forth in the introduction.

In the process of assessing the design results for validation of the methodologies reached at, it is proposed for future work to conduct a survey collecting empirical data through questionnaires, which may give rise to the creation of empirical models relating the properties of objects created with the judgmental impressions of parents and other relatives of the child-users. In this way, one may determine if the methodologies proposed and implemented giving rise to the design of toys are effective towards attaining the objectives that guided their creation.

It is also proposed for future studies to carry out a test phase with prototypes embodied in latex, since the prototypes shown were made in gypsum, thus postponing the confirmation and analysis of the properties and the strength of the material proposed (latex).

Acknowledgement

Parts of the research presented in this chapter were developed as part of the second author's Master of Science thesis in industrial design engineering, supervised by the first author.

Author details

Denis A. Coelho and Sónia A. Fernandes

Universidade da Beira Interior, Portugal

References

[1] Evaristo, Teresa (2006). Educação para a Cidadania, Guião de Educação para a Sustentabilidade-Carta da Terra' [in Portuguese- Citizenship Education, Guidelines for Education for Sustainability- Earth Charter], Ed. Ministério da Educação, 39 pp

[2] Fuad-Luke, Alastair (2004). 'The Eco-design Handbook', Thames & Hudson, 352 pp.

[3] Mustard, F. (2000). Early childhood development: The base for a learning society', Chapter presented at the HRDC or OECD Meeting, December 7, in Ottawa, Canada.

[4] Piaget, Jean and Inhelder, Barbel (1995). A Psicologia da Criança' [in Potuguese-'Citizenship Education, Guidelines for Education for Sustainability- Earth Charter], 2nd ed. Porto: Edições Asa

[5] Portuguese Ministry of Education- Ministério da Educação (2006). Educação para a Cidadania, Guião de Educação para a Sustentabilidade- Carta da Terra', Ed.: Ministério da Educação, Direcção-Geral de Inovação e de Desenvolvimento Curricular, 59 pp.

[6] Rutter, M. (2002). The interplay of nature, nurture and developmental influences: The challenge ahead for mental health', Archives of General Psychiatry 59, (11), 996-1000.

[7] Ryan, C. (2009). Design for Sustainability- a step-by-step Approach, Part II Chapter III A 'Quick-Start' approach to D4S, United Nations Environment Programme, Delft University of Technology.

[8] Wallon, Henri (1981). A evolução psicológica da criança' [in Portuguese- The Psychology of the Child], Lisboa: Edições 70, , 144.

Chapter 8

Sustainable Product Innovation: The Importance of the Front-End Stage in the Innovation Process

Kristel Dewulf

Additional information is available at the end of the chapter

1. Introduction

With an overpopulated planet, hungry for electricity and resources, sustainability will be one of the biggest challenges in the future. Present production and consumption patterns are causing serious environmental and human problems and cannot be sustained in a world with rising human aspirations. The challenges and opportunities for sustainable innovation are immense, and the time horizon is shrinking. Going green isn't just about saving the planet; it's about finding a dynamic equilibrium between human and natural systems, between saving the environment, making profit and enhancing all stakeholders' quality of life. Companies, designers and engineers can play an important role in this transition process towards a sustainable and smarter society with an improved 'quality of life'.

The very early phase in the product innovation process, the so-called front-end of innovation (FEI), is the stage of the innovation process where product strategy formulation, opportunity identification, idea generation, idea selection and concept development take place and decisions about new product development are taken [1]. These first phases in the engineering design process have the largest impact on the end result of the project [1, 2] and the highest payback to one's investments [1]. Accordingly, the front-end of innovation is often described as being the root of success for any company hoping to compete on the basis of innovation [2].

Notwithstanding the logic behind integrating sustainability in the early stages of an innovation process, in practice it is flawed. Front-end innovation is a hot research topic, but there is still little research done on its relationship to design for sustainability.

This chapter addresses the existing knowledge in the field of sustainable product innovation and its relation to the front-end of new product development. The research in this chapter aims to contribute to the understanding and implementing of sustainability in the early stages of an innovation process. A short overview of the used research method is presented in the first part. Secondly, the concept of the front-end of new product development is introduced by its different definitions. The section also describes the importance of the FEI, different types of innovation processes and the functions, activities and characteristics of the Front End. The third part looks to the concept of Sustainable Product Innovation (SPI) together with its drivers and barriers. Furthermore, it reflects on the current practice of the use of Sustainable Product Innovation tools. Next, the importance of integrating environmental considerations in the Front-End stage is presented. Different research results, insights and challenges are discussed in the penultimate part in order to identify successful patterns. At the end of this chapter, a summary is presented.

2. Research method

The represented research in this book chapter reviews the major works on the Front End of Innovation, Sustainable Design and the current state of the art in literature of front-end sustainability. It aims to identify gaps, challenges, issues and opportunities for further study and research. A literature review seems to be a valid approach, as it is a necessary step in structuring an in-depth research field and forms an integral part of any research conducted [3].

The main focus of this book chapter is the Product Innovation Process. Articles focusing on other aspects of an innovation process were not included, e.g. the review does not includenew marketing methods, or dimensions on new organizational methods in business practices, workplace organization or external relations. To limit the number of publications, papers mainly addressing sustainable design on a macro ecology level were also excluded from the review. Similarly, research with a highly craftsmanship rather than an industrial product design perspective were also excluded. Although these variables might be important antecedents to how firms eventually perform their FE activities, they are not focused on here.

Furthermore, the term denoting 'product' has several meanings frequently used in literature. In this book chapter, the term product means either the physical form of an object, a service or otherwise a product-service system.

3. Front End of innovation

It is important to understand the nature and outcomes of the Front End of Innovation before we can go deeper in the relationship between Sustainable Product Innovation and the early stages of an innovation process. In this section we give a short overview on the different aspects of the front end.

3.1. The Front End of new product development

The Front End (FE) is considered as the first stage of new product development, which roughly concerns the period from the idea generation to its approval for development, or its termination [4].Moenaert et al. [5] define the Front End as the process in which an organization formulates a product concept and decides whether or not to invest resources in that concept. Khurana and Rosenthal[6] note that the FE begins when an opportunity is first considered worthy of further ideation, exploration and assessment and ends when a firm decides to invest in the idea, commits significant resources to its development, and launch the project. The FE includes product strategy formulation and communication, opportunity identification and assessment, idea generation, product definition, project planning, and early executive reviews, which typically precede detailed design and development of a new product. One of the many other definitions of FE was formulated by Kim and Wilemon [7]; the Fuzzy Front End begins when an opportunity is first considered worthy of further ideation, exploration, and assessment and ends when a firm decides to invest in the idea, commit significant resources to its development, and launch the project' or shortly;the FE is the period between when a opportunity is first considered and when an idea is judged ready for development' [7].

Crawford and Di Benedetto[8] describe that the process in the FEI gives an answer to the primary questions: whether, what, why, who, when and how.

The decision is made whether or not a product innovation project passes to real development.

• What: the description of the project to be developed.

• Why: what is the strategy behind this new product development?

• Who: describes the human resources necessary to perform the development

• When: describes the timing of the project

• How: describes all the product requirements regarding the new development

We can detect some small variations in the above-mentioned explanations of the Front End. The definitions differ from author to author. Similar to Jacoby [9], we define the FE phase as 'all initial innovation activities, prior to development and ends where real new product development (NPD) starts'.

In contrast with new product development, there is no common terminology in academic literature and design practice as how to denote the early stages of an innovation process.

Different synonyms for the Front End can be found in literature; Fuzzy Front End [1], Front End of Innovation [10], pre-development [11], Phase zero, Stage zero, pre phase zero [12] or pre-project activities [13].

Cooper [11] introduced the term "pre-development" in 1988. Smith and Reinertsen first popularized the term "Fuzzy Front End" in 1991 [14]. Later on, in 1997, Verganti [12] described these pre-development activities as "the early stages of development" or the "pre-project activities" while Khurana & Rosenthal [13] used the term "pre-phase zero" in 1997.

Koen et al. [10] were the first to use the term Front End of Innovation in 2002, with the purpose of replacing the "Fuzzy Front End". The reasoning behind the wish to abandon the term Fuzzy Front End is that the word "fuzzy" implies that the Front End is mysterious, lacks accountability, and cannot be critically evaluated [15]. In this book chapter, we will refer to these early stages as the Front End of Innovation (FEI).

3.2. The importance of the Front End of innovation

The outcome of the FE process is of great importance on the innovation phases that come after the FEI. A variety of authors have recommended to focus/focusing on these early stages of NPD [16] [14] [9] [4] [13] [2] [15] [17]. This section gives an overview of different insights found in literature.

The complexity and cost of the complete innovation process depend to a large extent on the input: ideas for new products, user needs that have been detected, technological opportunities that have been scouted, choices that have been made between different options, and so on. Product success and firm success are to a large extent depending on decisions made in the FEI. The impact decisions can have on the final product result decreases along with the project evolution: whereas FEI decisions can impact the product as a whole, NPD decision have to take into account earlier decisions and can only have an impact on partial aspects of products [9].

Prior research by Khurana and Rosenthal [6] has pointed the importance of the early stages of the innovation process. Although an innovative company must be proficient in all phases of the new product development process, the most significant benefits can be achieved through improvements in the performance of the FE activities [6]. Also a study by Koen et al. [1] identified the front end as the key-contributing factor for new products. The FE presents one of the greatest opportunities for improving the overall innovation process [1]. Reid &Brentani focus on the roots of success; The FFE is the breeding ground for all new goods and services. Activities in the FFE are the root of success for any company hoping to compete on the basis of innovations [2]. Verworn [17] state that the best opportunities for improvement of the innovation process lie in the front-end activities. She suggests that a better understanding of the FEI, leads to a higher success rate in the overall new product development process. Koen and Bertels [15] highlight the importance of the path-dependency in an innovation process; the front end is very important because the product-development process is path dependent. This means that choices made in the front end lead to options as well as limitations regarding which products a company will ultimately be able to develop [15].

In Figure 1 the relationship is shown between influence, cost of change, and available information during the innovation process. At the beginning of the process, i.e. during the front-end, the degree of freedom and influence on the project outcome is high, while little information is available and the cost of changes is low. At later stages in the process one has more information available, but then the cost of change will increase. Decisions made in the front-end influence all subsequent phases of the innovation process. Quality, costs, and timings are mostly set during the front-end [18]. The challenge in the FE is created by the low amount of information and certainty. Once the specification for the future product is set at the FE, only relatively minor changes of the products are possible – or they will be very expensive and time consuming.

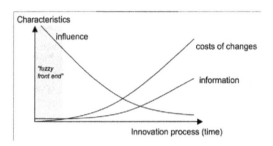

Figure 1. Evolution of influence, costs of changes, and information during the innovation process (von Hippel [19], modified by Herstatt & Verworn [18])

3.3. Front End functions

The reason of existence of the FE can be described through the functions of the FE. Jacoby [9] defines 6 crucial functions of the FE:

3.3.1. Product definition

The ultimate goal of the FE is to know what kind of product functions, product sub-functions and product characteristics a future product should have. It is not only about the idea of the new product or service, but also the added value it would represent and the major requirements [9]. As Cooper proposes [20], the desired goal of the FE is the creation of a well-defined product concept. A well defined product concept allows for a clearer understanding of development time, costs, required technical expertise, the right development team, market potential and positioning, risk, and organizational fit [4]. The product definition depends to large extend on the understanding of the customer needs, wants and preferences [21].

3.3.2. To define new business cases

The product definition and the business cases both describe the future business opportunity, but whereas the product definition focuses more on the product or the service itself, the business cases cover the possible benefits against the investments required. From that business point of view, a set of issues has to be addressed in the FEI in order to be able to complete a business case, such as business and product strategy, target market, product positioning, competition, marketing and finance [9].

3.3.3. Lower possible risks and reduce uncertainty

Uncovering possible pitfalls and reducing uncertainty is a crucial function of the FE. When a product or service is considered ready to enter the stage of NPD, it is important to know that this development project can be carried out with controllable risks and technological and market uncertainties [9]. Uncertainties also refer to the freedom of operations relating to existing patents and regulatory requirements [8] [22].

3.3.4. To decide on projects and products

Bringing a new product or service into development, means to make choices, to determine priorities and to allocate resources. Decision-making also refers to prioritizing between different products or projects. Go/no-go and prioritizing decisions not only takes place at the end of the entire FEI process but also during the different sub-phases within the FEI based on specific evaluation criteria [9]. Many of the tools used in FEI have the purpose to force decisions. Carbonell-Foulquié et al. [23] categorize the different criteria in five dimensions: strategic fit, technical feasibility, customer acceptance, market opportunity and financial performance.

3.3.5. To plan projects

Project planning frames a project in a certain time, usually with defined stages, objectives, deliverables and designated human and financial resources, next to all the other defining and decision activities.

3.3.6. To process and communicate information

Gathering & processing information or informing the organization is a key element in the FEI [9]. Processing information is a necessary condition for many of the other process functions described [4]. Moenaert et al. [5] give evidence of the fact that communication flows between organizational functions contribute to innovation success. In general terms, every analysis or synthesis in the FEI, one way or another, is based on processed information. These information flows could be very informal and tacit.

3.4. The Front End in an innovation process

Over the last two decades, several researchers and companies have suggested different approaches to innovation in the context of new product development. Many models can be found in literature and practice [24]. This section offers some background in product innovation models, with special attention to the Front End of the Innovation Process and the activities that occur/involved within these early stages.

Innovation can occur in many different areas of an organization. The 'Oslo Manual' produced by the OECD [25] defines innovation as: *'The implementation of a new or significantly improved product (good or service), or process, a new marketing method, or a new organizational method in business practices, workplace organization or external relations'*. We will focus in this section on the product development process.

Two notable pieces of work that have emerged from the FEI research are the New Concept Development Model (NCD) developed by Koen et al. [1] and Cooper's Stage-Gate process. The original Stage-Gate framework uses a sequential process with specified steps and timing, while Koen et al. presents a non-sequential relationship model. Both approaches and there relation to the FE will be explained more in detail in this section.

3.4.1. The new concept development model

Innovation projects in industry generally move along three major activity domains as shown in Figure 2.

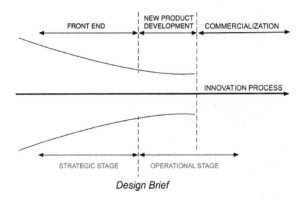

Figure 2. The innovation process. Adapted from Koen et al. [1]

The Front End of Innovation (FEI) or pre-development activities where future products are defined and decided on.

The New Product Development (NPD) where the products are actually developed.

The launching or commercialization activities where these newly developed products are brought to the market.

The NCD model shown in Figure 3 provides a good summary of the main FEI activities that occur prior to the Product Development Stage and consist of three parts: the relatively uncontrollable influencing factors, the engine that drives the activities of the FEI, and the five activity elements. These three key parts are explained more in detail below.

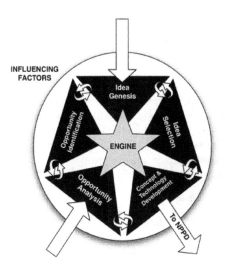

Figure 3. The New Concept Development (NCD) Model according to Koen et al. [1] (Image reproduced by the author)

The **'Engine'** represents the leadership, culture and business strategy of the organization that drives the five key elements.

The inner spoke area of the NCD model defines the **'Five Activity Elements'**:

Opportunity identification

In this element, large or incremental business and technological chances and opportunities are identified, by design or default, in a more or less structured way. According to Koen et al. [10], the sources and methods employed by the company can range from formal, systematic tools such as future scenario mapping or problem-solving methods such as the fishbone diagram as well as less formal, ad hoc approaches such as water-cooler conversations or individual insights.

Opportunity analysis

The second activity involves gathering together the additional information required in order to translate the identified opportunities into specific business and technology opportunities for the company. This may involve focus groups, market studies and/or scientific experiments. The level of effort put into these activities is dependent upon the attractiveness of the opportunity, the size of the future development effort, the fit with the business strategy and culture, and the risk tolerance of the decision makers.

Idea genesis

The third element is the idea genesis, which is described as the birth, development and maturation of the opportunity into a concrete idea. This represents an evolutionary and iterative process, including brainstorming sessions and idea banks, in which ideas are built upon, torn down, combined, reshaped, modified and upgraded. A new idea may emerge internally or come from outside inputs, e.g. a supplier offering a new material/technology or from a customer or user with a request.

Idea selection

Normally there are more opportunities and ideas than can be supported with the funding and time available within the company. The critical activity is to choose which ideas to pursue in order to achieve the most business and consumer value. The activity of prioritizing and selecting ideas may be based on an individual's choice or a comprehensive portfolio planning approach. Project selection, financial return and resource allocation in the FE is often just a wild guess, due to the limited information and understanding at this point.

Concept and technology development

Within this activity of the FE, the business case is developed, based on estimates of the other activities; market potential, customer needs, investment requirements, competition analysis and project uncertainty. This element is often seen as the final output of the FEI.

The relatively uncontrollable **'Influencing Factors'** consist of organizational capabilities, the outside world (distribution channels, law, government policy, customers, competitors, and political and economic climate) and the enabling sciences (internal and external) that may be involved. The 'influencing factors affect the decisions of the two inner parts.

The NCP model is a relationship model, not a linear process. The circular shape is meant to suggest that ideas are expected to flow and iterate between the five elements. Iteration and loop-backs are part of FE activities. The key elements of the FE are expected to proceed none-sequentially, as shown by the looping arrows between the elements. Interactions and intermingling between the influencing factors, the five key elements, and the engine are expected to occur continuously.

3.4.2. Stage-gate systems

Product innovation is a dynamic process; it begins with the discovery of new opportunities and product ideas and ends with the successful launch of a new product. Stage-Gate systems divide the steps between these point into a series of stages (=activities) and management decision gates. The original Stage-Gate model, introduced in the mid-1980s by Cooper, was based on research that focused on what successful project teams and businesses did when they developed wining products.

A Stage-Gate System provides a conceptual and operational road map to facilitate a project for moving a new-product project from idea to launch. It is a blueprint to improve effectiveness and efficiency [26]. The stages are where the work is done. Each gate serves as a Go/Kill/Hold/Recycle and prioritization decision point. Stage-Gate systems should provide a clear idea of where the project stands, where it is going, and what needs to be done next.

The typical Stage-Gate system is explained below and shown in Figure 4 [26].

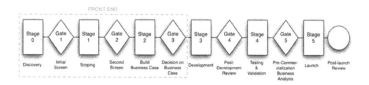

Figure 4. Overview of a Stage-Gate System. Adapted from [26]

Stage 0 – Idea / Discovery

Activities designed to discover opportunities and to generate new product ideas.

Stage 1- Scoping / Preliminary Assessment

A first quick and inexpensive assessment of the technical & marketplace merits of the project, so the project can be reevaluated more thoroughly at gate 2.

Stage 2 – Build Business Case

This is the final stage prior to product development. It is the stage that must verify the attractiveness of the project prior to heavy spending. And it is the stage where the project and product must be clearly defined. Here, market research, a detailed technical appraisal and a detailed financial analysis are undertaken.

Stage 3 – Development

Stage 3 involves the development of the product and of detailed test, marketing and operations plans. An updated financial analysis is prepared, and legal/patent/copyright issues are resolved.

Stage 4 – Testing & Validation

The purpose of this stage is to test the entire viability of the project: the product itself, the production/manufacturing process, customer acceptance, and the economics of the project.

Stage 5 – Launch

The final stage involves the full commercialization of the product; the implementation of both the marketing launch plan and the operations plan.

Post Launch Review / Post-Implementation Review

At some point, the product becomes a 'regular' product in the firm's line. This is the point where the project and product's performance is reviewed. A post-audit is carried out; the latest data on revenues, cost, expenditures, profits are analyzed together with a critical assessment of the project strengths and weaknesses, what we can learn from this project, and how we can do the next one better.

Preceding each stage is a decision point or gate. Gates are characterized by a list of pre-established criteria, ensuring that all projects are evaluated consistently and fairly. The role of the gatekeepers is to take a Go/Kill/Hold/Recycle decision and to review and approve the action plan for the next gate. Deliverables for the next gate must be clearly specified [26].

The standard 5-Stage, 5-Gate Stage-Gate New Product Process shown in Figure x is fairly generic. It serves as a sample or skeleton from which to develop a custom-tailored model. Not all projects pass through every stage of the model. Stage-gate processes are not rigid process steps and should be adapted to the context they are used in. None of the activities is mandatory – each project is unique [26]. The project leader considers what activities seem reasonable for the next stage. According to Jacoby, specific activities could belong both to the FEI and the NPD [9]. Also parallel processing is an important feature of stage-gate systems. Activities are parallel rather than sequential. Parallel processing compresses the development cycle without sacrificing quality [26]. Note that today's Stage-Gate processes are flexible, adaptive and scalable: they are iterative and features loops within these stages and potentially to previous stages [27].

Factors	Content
Factor 1 Strategic Fit and Importance	Alignment of project with our business's strategy. Importance of project to the strategy. Impact on the business.
Factor 2 Product and Competitive Advantage	Product delivers unique customer or user benefits. Product offers customer/user excellent value for money (compelling value proposition). Differentiated product in eyes of customer/user. Positive customer/user feedback on product concept (concept test results).
Factor 3 Market Attractiveness	Market size. Market growth and future potential. Margins earned by players in this market. Competitiveness - how tough and intense competition is (negative).
Factor 4 Core Competencies Leverage	Project leverages our core competencies and strengths in: technology, production/operations, marketing, and distribution/sales force.
Factor 5 Technical Feasibility	Size of technical gap (straightforward to do). Technical complexity (few barriers, solution envisioned). Familiarity of technology to our business. Technical results to date (proof of concept).
Factor 6 Financial Reward versus Risk	Size of financial opportunity. Financial return (NPV, ECV, IRR). Productivity Index (PI). Certainty of financial estimates. Level of risk and ability to address risks.

Table 1. Typical scorecard for Gate 3 [27]

In the next paragraph we will give some insights in the different activities and sub-phases of the FEI in the Stage-Gate model. A project cannot pass into the next stage until the evaluation is done and the gate is opened.

The front end is typically thought of as consisting of the first three sequential stages of the Stage-Gate process with the remaining stages focusing on the development process: discovery, scoping and building a business case, as also shown in Figure 4. The decision to "move into a full-scale development project" cannot be taken until the Gate 3 criteria have been met. In the early stages, these criteria tend to be largely qualitative and deal with 'must meet' and 'should meet' criteria [26]. Scorecards are based on the premise that qualitative criteria are often better predictions of success than financial projections. In use, management develops a list of about 6-8 key criteria, know predictors of success, on a scorecard. A typical scorecard for Gate 3 is present in Table 1. Note that different scorecards and criteria are used for different types of projects.

3.4.3. Conclusions

Two notable pieces of work that have emerged from the FEI research were presented in this section; Cooper's original sequential Stage-Gate process with specified steps and timing, and the non-sequential NCD relationship model from Koen et al. From a sustainable product innovation perspective, the problem with these frameworks is that they do not explicitly explain how sustainable design considerations can be integrated into the front end. None of them mentioned sustainability or provide sustainable design guidelines in the presented methodology.

3.4.4. Front End activities

There is no such a thing as a universal set of activities necessary to the FEI. The description of the different pre-development activities differs from author to author. The required innovation level, the context of the company, the available time, resources, strategy and markets...will usually determine the set of activities. A summary of the activities found throughout literature was made by Jacoby [9] and is presented in Table 2.

3.5. Front End characteristics

The FE phase is fundamentally different from the development stage of the innovation process. Characteristics of the FE compared to the traditional development phase is summarized in Table 3. Though all innovation processes does not follow a single pattern, the FE phase is intrinsically non-routine, dynamic and uncertain [7]. Essentially, the front end requires more expansive and divergent thinking [15]. The ambiguity level at the end of the FE can affect the risk related to the identified idea in the development stage [7].

Activity in the FEI	Author
Idea generation Product concept Idea genesis Business ideas Concept development	Cooper (1994), Murphy & Kumar (1997), Khurana& Rosenthal (1998), Montoya-Weiss &O'Driscoll (2000), Koen et al. (2001), Krishnan & Ulrich (2001), Nobelius&Trygg (2002), Langerak et al. (2004), Sandmeier et al. (2004), Buijs&Valkenburg (2005), Braet&Verhaert (2007)
Idea qualification Idea selection Idea screening Concept screening	Montoya-Weiss &O'Driscoll (2000), Koen et al. (2001), Nobelius&Trygg (2002), Langerak et al. (2004), Sandmeier et al. (2004), Cooper (2008), Braet&Verhaert (2007)
Opportunity identification & analysis Discovery	Koen et al. (2001), Cooper (2008)
Search areas Scoping	Buijs&Valkenburg (2005), Cooper (2008)
Product definition Design brief Concept definition	Cooper (1988), Murphy & Kumar (1997), Khurana& Rosenthal (1998), Nobelius&Trygg (2002), Buijs&Valkenburg (2005), Braet&Verhaert (2007)
Project evaluation	Cooper (1988), Murphy & Kumar (1997)
Product & strategic planning Product & portfolio strategy	Verganti (1997), Khurana& Rosenthal (1998), Langerak et al. (2004), Crawford (2006)
Concept generation	Verganti (1997), Koen et al. (2001), Crawford (2006)
Pre-technical evaluation	Cooper (1994)
(Preliminary) investigation	Cooper (1994), Verganti (1997)
Building business case (or value) Business plan concept	Cooper (1994), Hughes & Chaffin (1996), Sandmeier et al. (2004), Cooper (2008)
Business analysis	Nobelius&Trygg (2002), Langerak et al. (2004)
Capture market value Market analysis Market opportunities	Hughes & Chaffin (1996), Khurana& Rosenthal (1998), Sandmeier et al.(2004)
Technological analysis Technological opportunities	Khurana& Rosenthal (1998), Sandmeier et al. (2004)
Deliver winning solution	Hughes & Chaffin (1996)
Project & process planning	Hughes & Chaffin (1996), Khurana& Rosenthal (1998), Nobelius&Trygg(2002)
Feasibility	Khurana& Rosenthal (1998)

Table 2. Activities in the FEI according to different authors [9]

	Front End	Product Development
Nature of work /	Experimental	Disciplined
Method	Often chaotic	Structured
	'Eureka' moments	Systematic
	Often unstructured	Goal-oriented with project plan
	Creative	
Degree of formalization	Low	High
Activity	Individuals and team conducting research to	Multifunction product and/or process-
	minimize risk and optimize potential	development team
State of an idea	Probable	Determined
	Often fuzzy	Clear
	Easy to change	Specific
	Easy to reject	Difficult to change
		More difficult to reject
Information	Often qualitative	Quantitative
	Informal	Formal
	Approximate	Precise
Degree of formalization	Low	High
Personnel involvement	Individual or small project team	Full development team
Commitment of the CEO	None or small	Usually high
Funding	Variable.	Budgeted
	In the beginning phases many projects may	
	be 'boot legged', while others will need	
	funding to proceed.	
Revenue expectations	Often uncertain	Predictable with increasing certainty analysis
	Great deal of speculation	and documentation as the product release
		date gets closer
Damage if abandoned	Usually small	Substantial
Commercialization date	Uncertain or unpredictable	High degree of certainty
Measure of progress	Strengthened concepts	Milestone achievement

Table 3. Difference between Front End of Innovation and the New-Product development process. Based on Koen et al. [10] and Kim & Willemon

Figure 5 below shows that the 'fuzziness level' of an idea gradually diminishes as the NPD process progresses [7]. When the fuzziness level or uncertainty level descends below the 're-quired' approval level (a) for a specific firm, the development phase usually begins. The start of development phase is the intersection point (b). The ambiguity level at the end of the FE can affect the risk related to the identified idea in the development phase. The approval decision at the end of the FE is usually the first formal go/no-go decision. It is a critical point, as it determines whether the firm will invest and if so, how much budget, time, people… it is willing to invest.

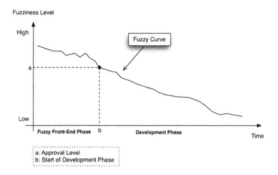

Figure 5. Patterns of the fuzziness level through the New Product Development [7] (Image reproduced by the author)

3.6. Conclusions

High-quality up-front analysis is essential to effective and efficient product development. Various authors have pointed out the importance of the Front End of Innovation. Over de last decade, the number of publications on the Front End has increased, providing more insights on the characteristics, process, activities, functions, and patterns in the Front End.

However, from a sustainable product innovation perspective, the Front End literature does not explicitly explain how sustainable design considerations can be integrated into these early stages. This problem is discussed further in Section 5.

4. Sustainable product design

4.1. What is sustainable product design?

4.1.1. Sustainable development

The World Commission on Environment and Development defined Sustainable Development in 1987 as 'A development that meets the needs of the present without compromising the ability of future generations to meet their needs' [28]. This definition has been taken over, reformulated and evolved over time by many. Other definitions on sustainability focus on the so-called 'triple bottom line': the three dimensions people, planet, and profit, also called social equity, economic efficiency and environmental performance.The International Institute for Sustainable Development in conjunction with the World Business Council for Sustainable Development has defined sustainable development from a business perspective view. "Sustainable development means adopting business strategies and activities that meet the needs of the enterprise and its stakeholders today while protecting, sustaining and enhancing the human and natural resources that will be needed in the future" [29]. The stakeholders include shareholders, lenders, customers, employees, suppliers and communities who are affected by the organization's activities. This definition also highlights business's dependence on human and natural resources, in addition to physical and financial capital. The book "Sustainability by Design" by John Ehrenfeld is founded upon a new definition "Sustainability is the possibility that humans and other life will flourish on earth forever" [30].

4.1.2. Ecodesign

Ecodesign and Design for Environment (DfE) are terms for strategies that aim to integrate environmental considerations into product design and development. They involve life-cycle thinking, which means the integration of life-cycle considerations into product design. The overall goal is to minimize the consumption of natural resources and energy and the consequent impact on the environment while maximizing the benefits for customers [31].

The European Parlement formulated one of the many other definitions of Ecodesign in 2005 in Directive 2005/32/EG. 'Eco-design means the integration of environmental aspects into product design with the aim of improving the environmental performance of the product throughout its whole life cycle'. There are lots of synonyms for the term Ecodesign. Bhander et al. [32] uses the following synonyms: Design for Environment (DfE), Eco-Design, Eco-innovation, Environmentally Conscious Design (ECD) and Sustainable Design. According to O'Hare [33] ECD is the umbrella term for eco-design, eco-innovation and DfE. Sustainable Design is any form of ECD that affects a social and an economic aspect as well as the ecological aspect.

4.1.3. Sustainable design

Sustainable product design (SPD) is more than Ecodesign, as it integrates social and ethical aspects of the product's life cycle alongside environmental and economic considerations, aiming for the so-called 'triple bottom line'. Sustainable product development and design is concerned with balancing economic, environmental and social aspects in the creation of products and services [31]. A framework that shows the relationship between the different disciplines is shown in Figure 6.

Figure 6. Relationship between ecodesign, sustainable design and sustainable development [31]

McLennan [34] defines Sustainable Design as 'a design philosophy that seeks to minimize or eliminate negative impact to the natural environment through skillful, sensitive design'.

Up till now, most attention has been given to environmental sustainability within the design process, both by academics as practitioners [35].

4.2. Drivers and barriers for sustainable design

Usually, there's no one-single driver or barrier for Sustainable Design. A combination of several factors, both internal and external determines whether a firm chooses the path of sustainable design or not. For this, the firm policy needs to find a balance between environmental, social and economic needs. This section summarizes the main findings in literature on the stimuli and barriers for enterprises to practice sustainable product design.

4.2.1. Drivers for sustainable design

Extensive research on drivers for sustainable design has been carried out by research organizations and industrial companies in the last two decades with significant insights achieved [36] [31] [37] [38]. Charter and Tischner mentioned already in 2001 a growing number of drivers for sustainable design worldwide. [31]

Van Hemel and Cramer [38] did a study on stimuli and barriers for ecodesign in SMEs and listed the most common and influential factors. The most influential internal stimuli were the opportunities for innovation, the expected increase of product quality and the potential market opportunities. The research revealed quite clearly that the most influential external stimuli for ecodesign are 'Customer demands', 'Governmental legislation' and 'Industrial sector initiatives'.

A recent study based on a survey conducted on 10.000 multidisciplinary professionals from Spanish innovation driven companies [37] shows that sustainability is a cardinal driver for innovation. Moreover, the study indicates that the main drivers for integrating sustainable criteria are environmental impact reduction, energy efficiency, marketing and brand value, and legislation adjustment. Internal drivers for practicing sustainable design can vary depending on the size of the company. The obtained results in the study show that for micro companies (1-10 employees), the driving element most considered is cost reduction, followed by marketing and brand value. On the other hand the driving elements less considered for micro companies are legislation adjustment and to avoid economic sanctions. In the case of SMEs (10-250 employees) the main driver is the client demand and being fashionable and the least important is to avoid economic sanctions. Finally for macro companies (over 250 employees) the least important driver is cost reduction. The most important ones are to avoid economic sanctions and legislation adjustment. Nowadays there are a lot of EU directives on a wide range of categories. These EU regulations determine common rules concerning responsibility and technical issues such as the end-of-life disposal treatment of equipment and financial issues for instance who has to pay for disposal treatment. By all the directive requirements, companies are forced to constantly push their limits concerning sustainable development.

According to Mathieux et al. [39] there are also some parallel benefits regarded from business perspective if a firm decides to practice sustainable design. The product design team is stimulated to see the bigger picture, as they need to make decisions based on life-cycle thinking. This can give a greater understanding in the complex chain of stakeholders of the company and acquire a global view of the market opportunities, cost saving and the product portfolio.

4.2.2. Barriers for sustainable design

Not all companies chose to practice sustainable design. Some of them are struggling to integrate this way of thinking into their current design process, portfolio and business structure. The main barriers found in literature are listed below.

First of all, the board needs to be convinced of the goal they're setting. If there's a management's lack of commitment, as stated in [38], [33], [37] or if environmental improvements are perceived as not their responsibility, then it's practically impossible to implement sustainable design in an enterprise. The fact that a firm sees no clear environmental benefit is also often mentioned as a very important barrier [38]. Another common reason is the lack of acquisition of tangible benefits. This refers to the absence of direct benefits in the short term, such as the growth of production or sales, fiscal incentives or client satisfaction [37].

Also the practitioners need to be convinced of the new method that they're going to implement. The design team needs to be sure they can benefit from the ecological conscious design (ECD) tools. It is possible that a new tool isn't useful from the first second, tools often requires patience and has to be customized to the specific need of the design team [33]. On top of all this, the design team has to have the attitude, appropriate knowledge and skills to design and develop sustainable products [40].

Another frequent obstacle is the fact that an ecological optimized product can be in conflict with its functional product requirement. It can be a challenge for the designers and design teams to integrate the sustainable requirements without compromising on the technical possibilities and the functional needs.

For many companies, cost is a very important element when taking environmental oriented decisions [37]. Not only the costs for the optimization of the product need to be taken into account. Also the general costs to create an environment in a firm where it is possible to practice sustainable design. Maybe there is the need for a new team, including experts of the environmental or ecological sector, or a new structure, a new vision or an adjusted view on the business model. All this may cause a significant augmentation of the overheads. And that is what scares some companies to make the switch to a more environmental policy. Moreover, if the general cost of an enterprise rise, there is the possibility they can't compete with the direct competitors. This may lead to a commercial disadvantage.

One of the most important external barriers is the lack of involvement of consumers [37]. If they are not willing to pay for it, or simply aren't interested in an environmental friendly product, the whole project is doomed to fail. In general, market demand steers the companies whether to choose for sustainable design. If in a certain product sector the demand for environmental products rises, the entire sector will develop toward these kinds of products. The other way around is also possible; companies will hesitate to implement sustainable design if the market shows only little interest in these kinds of products. The influence of costumers can be very decisive for Sustainable Design.

From the study from van Hemel and Cramer [38] can furthermore be derived that three barriers must be characterized as 'no-go' barriers; their existence obstructs the ecodesign improvement options in question from being implemented. These were the following barriers: 'No clear environmental benefit', 'Not perceived as responsibility' and 'No alternative solution is available'.

4.2.3. Summary

The main drivers and barriers found in literature to practice sustainable design are summarized in Table 4. A distinction in the table is made between the forces within and outside the firm that gives the motivation whether or not to incorporate sustainability criteria in products.

Internal drivers and barriers are the internal factors that originate inside the company itself. External stimuli and barriers are the external factors that influence the decisions made towards sustainable design from outside the company.

In conclusion, practicing sustainable design is balancing between all the mentioned drivers and barriers. Every company needs to consider and determine its own specific requirements and goals.

Enhancing sustainable design does not only depend on finding alternative solutions for technical problems. Even more important are economical and social factors like the acceptance of environmentally improved products in the market, and the way the market will perceive these products. Sustainable design is most successful when supported by several strong internal and external stimuli and not blocked by any no-go barriers. It only stands a chance, if it is supported by stimuli other than the expected environmental benefit alone. Contrary to prevailing literature on environmental management in SMEs, Van Hemel and Cramer [38] concluded in their study that internal stimuli are a stronger driving force for ecodesign than external stimuli.

Many drivers and barriers for Sustainable Design have their roots in the Front End of Innovation. This topic will be discussed in the next section.

	Drivers	Barriers
Internal	Management's sense of responsibility	No clear or lack of environmental benefits
	Business opportunities	Lack of acquisition of tangible benefits
	Innovational opportunities	Conflict with functional product requirements
	Risk Management	Not perceived as responsibility
	Long-term survival	No alternative solution available
	Competitive advantage	Lack of management's commitment
	Improvement of product and product quality	Extra costs
	Improvement of brand image	Shortage of short-term benefits
	Cost reduction	Lack of understanding of sustainable design tools
	Environmental impact reduction	Lack of acquisition of tangible benefits
	Energy efficiency	
	The need for innovative power	
External	Governmental Regulations / Legislation	Lack of interest from consumers / Lack of market demand
	Increase awareness of the public / Public pressure	Consumers not willing to pay (extra) for it
	Customer needs and demands	Commercial disadvantage
	Growing pressure from different stakeholders	
	Market competition / Being 'fashionable'	
	New market opportunities	
	Cooperation with supply chain partners	
	Development of external assessments (labels, standards…)	
	Availability of subsidies	
	Growing amount of knowledge	
	Industrial sector initiatives	

Table 4. Drivers & Barriers for Sustainable Design

5. Conclusions and insights on the importance of integrating sustainability in the Front End

High-quality up-front analysis is essential to effective and efficient product development. Various authors have pointed out the importance of the Front End of Innovation. Over de last decade, the number of publications on the Front End has increased, providing more insights on the characteristics, process, activities, functions, and patterns in the Front End. However, from a sustainable product innovation perspective, the Front End literature does not explicitly explain how sustainable design considerations can be integrated into these early stages.

Notwithstanding the logic behind integrating sustainability in the early stages of an innovation process, in practice it is flawed. Front-end innovation is a hot research topic, but there is still little research done on its relationship to design for sustainability. There are a number of tools available to guide designers, engineers and managers in the design process when the specifications of the product or service are already set. However, methods supporting target identification for sustainable innovations are rare [41].

In light of the increasing attention to sustainability, sustainable product innovation and pre-development activities in new product development, various authors have recently pointed out the importance of integrating sustainability in the front end [41], [42], [43], [33], [44], [31], [45].

This sections aims to give an overview of the main reasons why integrating sustainability in the front end is so important.

5.1. Tackling sustainability problems at higher system levels

The international research literature on Sustainable Product Development (SPD) identifies the need to move beyond incremental change (e.g. redesign of existing products) to more fundamental, systematic changes. These are described as 'function innovation' or 'system innovation' [46].

Brezet's model of 'eco-design innovation' [46] defines four types of environmental innovation, characterized by product improvement, product redesign, function innovation and system level innovation, according to the environmental impact reduction or eco-efficiency that can be achieved, shown in Figure 7.

The vertical axis expresses the eco-efficiency or environmental impact improvement. For example, factor 2 equates to half the overall environmental impact of a product, or a factor 2 performance improvement in material and energy efficiency. The horizontal axis corresponds with the time that a company or industry needs to progress through on the way to achieving environmental sustainability.

The first two stages of 'product improvement' and 'redesign product' focus on lower systems levels and deliver small to moderate improvements in environmental sustainability.

The latter stages focus on function and system innovation and deliver considerably greater system improvements in environmental performance.

Modest eco-efficiency gains can be achieved with relatively little effort in the new product development stage of an innovation process. However, to tackle problems at higher system levels, the problem needs to be already integrated in the Front End.At a later stage in the innovation process, the design space is limited and the resources allocated, as also pointed out in the previous chapters. After this crucial phase only incremental environmental improvements or product redesign are possible.

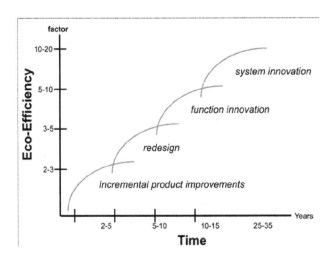

Figure 7. The four generic levels of eco-design, after Brezet [36]

5.2. Greening the design brief

A design brief is a written description of a project that requires some form of design. It is an agreement, or contract between the parties involved in the project. Often times, it is also a point of transfer between different professionals, where the project is handed over from marketing to design, or from a product manager to an in-house design team or external design agency. It is also a roadmap and project-tracking tool, defining the various steps that will be followed [47].

The role of a design brief is to provide the foundation to the entire design process and can be seen as the report or summary of the investigations steps and the decisions taken in the Front End, as shown in Figure 8 [48].

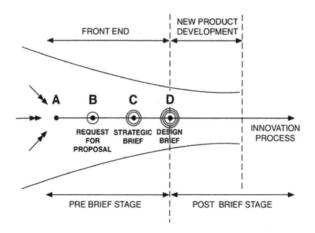

Figure 8. The various stages of a design brief in the front end of the innovation process [48]

Sustainable design projects would be far more effective if commencing from an environmentally responsible design brief. The secret to sustainable products lays upstream in the innovation process, in a good brief providing guidance to the design, engineering, and marketing and management team.

5.3. Commitment and allocation of resources

Decisions made in the front-end have a significant influence on all subsequent phases of the innovation process. For example, quality, costs, and timings are mostly set during the front-end phase [18]. The final approval at the end of the FE is usually a formal go/no-go decision. It is a critical point, as it determines whether the firm will invest and if so, how much it is willing to invest. This is also the moment where the other resources, e.g. time and people are allocated.

It is also a critical point regarding fruitful opportunities towards sustainable design, as the success of the product's final sustainability, is highly dependent on the previous committed resources. If no sufficient resources, e.g. time, budget and people with the right skills and knowledge are committed, the overall sustainability success is doomed to fail. Dewulf et al. [48] note that defining quantitative environmental targets in the early stage of an innovationproject often appears to be very difficult for innovation projects with a high innovation level. As costs and timings are mostly set during the front-end, this is a hard part to deal with.

5.4. Early tackling of barriers

Table 3 in Section 4.2.3 summarizes the internal and external drivers and barriers for sustainable design. A lot of those barriers have their roots in the Front-End. Performing the right actions in the FE can tackle various barriers. Performing an early financial and environmental analysis can, for example, make the benefits clear. One of other the barriers is lack of understanding of sustainability and sustainable design tools. To tackle this in an early stage, the intelligence level on those domains can to be increased by proper education and by providing relevant and reliable information.

5.5. Front-loading

In the FE stage, the degrees of freedom and influences on the project outcome are high, while little information is available and the cost of changes is low, as shown in Figure x. At later stages in the process one has more information available, but then the cost of change will increase. It is under these conditions that the front-end team needs to make decisions. That's why dealing with sustainability in the front-end of a product innovation process is often called 'wicked'; multidimensional with a complex interdependency. One promising method to deal with this 'wicked aspects' is 'front-loading'. Front-loading is defined as "a strategy that seeks to improve development performance by shifting the identification and solving of problems to earlier phases of a product development process" [49]. By spending more energy in the front phase on environmental analysis and strategic design one gets more information while the influence is high and the cost of change is low.

5.6. Doing the right thing vs. doing things right

Successful sustainable design requires both strategic (front-end) and operational (new product development) activities [50]. This perspective highlights the importance of including sustainability aspects already into the front-end activities of the innovation process in order for them to be considered at a strategic level [41]. The operational level is all about eco-efficiency or doing the things right, while the strategic level focuses on eco-efficiency or doing the right thing. Unfortunately, O'Hare [33] has noted that there is a lack of tools to support the early activities of eco-design in general. The majority of sustainability tools take the existing solution as a starting point, rather than considering the problem at a higher system level. They are generally intended for use after the strategic and conceptual design phase and cannot support the full range of challenges that are likely to be encountered during the front-end stage.

The study in this paper is part of a larger research at Delft University of Technology that is focusing on front-end sustainable product innovation. New research studies are planned to answer the question on how to best integrate sustainability in the front end. The insights provide by this book chapter will serve as the basis for future research.

Acknowledgements

The author would like to acknowledge the Howest Industrial Design Center in Belgium for their financial support for this research, and the Design for Sustainability research group at

Delft University of Technology, in particular dr. R. Wever and Prof. dr. J.C. Brezet, for their support and critical feedback.

Author details

Kristel Dewulf[1,2,3]

Address all correspondence to: kristel.dewulf@howest.be

1 Howest Industrial Design Center, Kortrijk, Belgium

2 Delft University of Technology (TUDelft), Faculty of Industrial Design Engineering, Section Design for Sustainability, Delft, CE Delft, The Netherlands

3 Ghent University (UGent), Faculty of Economics and Business Administration, Department Management, Innovation and Entrepreneurship, Research Group Corporate Social Responsibility, Ghent, Belgium

References

[1] Koen P, Ajamian G, Burkart R, Clamen A, Davidson J, D'Amore R, et al. Providing clarity and a common language to the "fuzzy front end". Research-Technology Management. 2001;44(2):46-55.

[2] Reid SE, De Brentani U. The Fuzzy Front End of New Product Development for Discontinuous Innovations: A Theoretical Model. The journal of Product Innovation Managment. 2004; 21, (3):170–84.

[3] Easterby-Smith M, Thorpe R, Lowe A. Management Research - An Introduction. London: Sage Publications; 2002.

[4] Murphy SA, Kumar V. The Front End of new product development: a Canadian survey. R&D Management. 1997 (27):5-15.

[5] Moenaert R, De Meyer A, Souder W, Deschoolmeester D. R&D / Marketing communication during the fuzzy front-end. IEEE Transactions on Engineering Management. 1995 (42):243-58.

[6] Khurana A, Rosenthal S. Towards Holistic "Front Ends" in New Product Development. The journal of Product Innovation Management. 1998 (15):57-74.

[7] Kim J, Wilemon D. Focusing the fuzzy front-end in new product development. R&D Management. 2002;32(4):269-79.

[8] Crawford CM, Di Benedetto CA. New Products Management. 8th edition ed. Homewood: McGraw-Hill/Irwin; 2005.

[9] Jacoby A. Performance in the Front-end of Innovation: linking strategy to requirements. Antwerp: University of Antwerp; 2012.

[10] Koen P, Ajamian G, Boyce S, Clamen A, Fischer E, Fountoulakis S, et al. Fuzzy-Front End: Effective Methods, Tools and Techniques. In: Belliveau P, Griffin A, Somermeyer S, editors. The PDMA ToolBook 1 for New Product Development: Wiley; 2002. p. 480.

[11] Cooper RG. Predevelopment activities determine New Product success. Industrial Marketing Management. 1988;17(3):237-47.

[12] Verganti R. Leveraging on systematic learning to manage the early phases of product innovation projects. R&D Management. 1997;27(4):377-92.

[13] Khurana A, Rosenthal S. Integrating the Fuzzy Front End of New Product Development. MIT Sloan Management Review. 1997:103-20.

[14] Smith P G RDG. Shortening the product development cycle. Research-Technology Management. 1992 (35):44-9.

[15] Koen P, Bertels H. Front End of Innovation. In: Malhotra JSN, editor. Wiley International Encyclopedia of Marketing. Volume 5: Product Innovation and Management: Wiley-Blackwell; 2011.

[16] Cooper RG. Third-Generation New Product Processes. The journal of Product Innovation Management. 1994;11(1):3-14.

[17] Verworn B. A structural equation model of the impact of the "fuzzy front end" on the success of new product development. Research Policy. 2009 Dec;38(10):1571-81. PubMed PMID: ISI:000272759500006.

[18] Herstatt C, Verworn B. The fuzzy front end of innovation. 2001; Working Papers / Technologie-und Innovationsmanagement, Technische Universität Hamburg-Harburg, No. 4.

[19] von Hippel E. Wettbewerbsfactor Zeit. Moderne Industrie; 1993; in The Fuzzy Front End of Innovation, working paper by Herstatt C. and Verworn, B., 2001.

[20] Cooper RG, Kleinschmidt EJ. Screening new products for potential winners. Long Range Planning. 1993;26(6):74-83.

[21] Buijs J, Valkenburg R. Integrale Productontwikkeling (in Dutch). Den Haag: Lemma; 2005.

[22] Braet J, Verhaert P. The practice of new products and new business. Leuven: Acco; 2006.

[23] Carbonell-Foulquié P, Munuera-Aléman J L, Rodriguez-EscuderoA I. Criteria em-
      ployed for go/no-go decisions when developing successful highly innovative prod-
      ucts. Industrial Marketing Management. 2004 (33):307-16.

[24] Dubberly H. How do you design? A Compendium of Models.:Dubberly Design Of-
      fice; 2004.

[25] OECD, Eurostat. Paris: OECD, Eurostat, 2005.

[26] Cooper RG. Stage-Gate Systems: A New Tools for Managing New Products. Business
      Horizons. 1990;33(3):44-54.

[27] Cooper RG. How companies are reinventing their idea-to-launch methodologies. Re-
      search Technology Management. 2009;52(2):47-57.

[28] Our Common Future: Report of the World Commission on Environment and Devel-
      opment. United Nations, 1987.

[29] Business strategy for sustainable development: leadership and accountability for the
      90s: International Institute for Sustainable Development in conjunction with Deloitte
      &Touche and the World Business Council for Sustainable Development; 1992.

[30] Ehrenfeld J. Sustainability by design : a subversive strategy for transforming our con-
      sumer culture. New Haven: Yale University Press; 2008. xxii, 246 p. p.

[31] Charter M, Tischner U. Sustainable solutions : developing products and services for
      the future. Sheffield: Greenleaf; 2001. 469 p. p.

[32] Bhander GS, Hauschild M, McAloone T. Implementing life cycle assessment in prod-
      uct development. Environmental Progress. 2003;22(4):255-67.

[33] O'Hare JA. Eco-Innovation tools for the early stages: an industry-based investigation
      of tool customisation and introduction. Bath, UK: University of Bath; 2010.

[34] McLennan JF. The philosophy of sustainable design : the future of architecture. Kan-
      sas City, Mo.: EcoTone ; [London : Publishing Services, distributor]; 2004.

[35] Suttclife LFR, Maier AM, Moultrie J, Clarkson PJ, editors. Development of a frame-
      work for assessing sustainability in new product development. 17th International
      Conference on Engineering Design (ICED'09); Stanford.

[36] Brezet H, van Hemel C. Ecodesign : a promising approach to sustainable production
      and consumption. Paris, France: UNEP; 1997.

[37] Santolaria M, Oliver-Sola J, Gasol CM, Morales-Pinzon T, Rieradevall J. Eco-design in
      innovation driven companies: perception, predictions and the main drivers of inte-
      gration. The Spanish example. Journal of Cleaner Production. 2011;19(12):1315-23.

[38] van Hemel C, Cramer J. Barriers and stimuli for ecodesign in SMEs. Journal of Clean-
      er Production. 2002;10(5):439-53.

[39] Mathieux F, Rebitzer G, Ferrendier S, Simon M, Froelich D. Ecodesign in the Europe-an Electr(on)ics Industry – An analysis of the current practices based on case studies. The Journal of Sustainable Product Design. 2001;1(4):233-45.

[40] Ammenberg J, Sundin E. Products in environmental management systems: drivers, barriers and experiences. Journal of Cleaner Production. 2005;13(4):405-15.

[41] Hassi L, Peck D, Dewulf K, Wever R. Sustainable Innovation, Organization and Goal Finding. Joint Actions on Climate Change; 2009; Aalborg, Denmark.

[42] Wever R, Boks C. Design for Sustainability in the Fuzzy Front End. Sustainable Inno-vation '07; 2007; Farnham, UK, October 29-30.

[43] Wever R, Boks C, Bakker C. Sustainability within Product Portfolio Management. Sustainable Innovation '08; 2008; Malmö, Sweden, October 27-28, (pp. 219-227)

[44] Petala E, Wever R, Dutilh C, Brezet H. The role of new product development briefs in implementing sustainability: a case study. Journal of Engineering and Technology Managment. 2010 (27):172-82.

[45] Hassi L, Wever R. Practices of a 'Green' Front End. A gateway to environmentel inno-vation. ERSCP-EMSU, Knowledge Collaboration & Learning for Sustainable Innova-tion; Delft, The Netherlands 2010.

[46] Brezet H. Dynamics in EcoDesign Practice. Industry and Environment. 1997;20(1-2): 21-4.

[47] Phillips PL. Creating the perfect design brief : how to manage design for strategic ad-vantage. New York: Allworth : Design Management Institute ; [Garsington : Wind-sor, distributor]; 2004. xvi, 191 p. p.

[48] Dewulf K, Wever R, Brezet H. Greening the Design Brief. In: Mitsutaka M, Yasushi U, Keijiro M, Shinchichi F, editors. Design for innovative value towards a Sustainable Society: SpringerLink; 2012. p. 457-62.

[49] Thomke S, Fujimoto T. The effect of front-loading proble-solving on product devel-opment performance. The Journal of Product Innnovation Management. 2000;17(2): 128-42.

[50] Ölundh G, Ritzin S. Making an Ecodesign Choice in Project Portfolio Selection. Inter-national Engineering Management Conference; 2004; Singapore, Republic of Singa-pore, October 18-21.

Case Studies

Chapter 9

Early Stages of Industrial Design Careers

Inalda A. L. L. M. Rodrigues and Denis A. Coelho

Additional information is available at the end of the chapter

1. Introduction

This chapter seeks to contribute to the elucidation of, not only, industrial designers, but also of educational institutions, both domestic and international, about how important it is to follow the development of markets and technology, in order to incorporate this in the curricula of future designers, with a view to further aid their transition to the labour market. The methodology set for the completion of the work was divided into two stages. Initially, literature review was used to underpin research and then empirical work was carried out in the search for new information and clarifications from the world of praxis.

The primary data collection was carried out using data collection instruments, which consisted of two questionnaires, applied to two distinct groups of respondents, namely, industrial designers and employers. The application of a survey instrument was chosen because of its ability to enable distance of the researchers to the respondenys and more autonomy than, for example, interviews (Almeida and Pinto, 1995). Questionnaires free the researcher from presence at the time of the response by the respondent, enabling an indirect interaction (Carmo and Ferreira, 1998). The dissemination of questionnaires to industrial designers was performed with the support of the Portuguese Design Centre, which contributed to the high number of respondents achieved (141).

1.1. Scope of the research

As this chapter reports on a study on the prospects for job market integration of industrial designers, it was necessary to point the investigation in a first phase to the jobs that are available in countries like Portugal, Germany, France, Italy, UK, Canada, Spain, United States of America, Macau, Brazil, East Timor, Guinea Bissau, São Tomé and Principe, Mozambique, Angola and Cape Verde, in order to unveil the true extent of demand for the profession of industrial designer in the Lusophone (Portuguese Speaking) Space, and Western Europe and the United States and Canada. The authors deliberately chose to focus this study on entrepreneurship and its importance for the industrial designer, as well as on the contents of training that are more valued by employers of these professionals. The authors chose to present the results of the survey of businesses employing industrial designers and of industrial designers in the final parts of the chapter. The focus of the chapter lies also to a greater extent on the first phase of the career of industrial designers, rather than on later stages, where the range of career paths is more diversified and more difficult to cover in a work of this nature. It is also believed that the initial phase of the designer's career is a critical stage for the affirmation of the individual professional and therefore it arouses a very high level of interest as an object of research.

2. Opportunities for integration of industrial designers in working life

In pursuing the proposed aims, it was necessary to define criteria for the selection of job advertisements that would be examined. Given the impossibility of collecting all the available advertisements, during the time in which the study took place, because of insufficient resources, information was collected only in the form of sampling. Advertisements were retrieved by searching the Internet, specifically those clearly aimed at industrial, product, or equipment designers and that provided some information on sector of work, on the required qualifications, and on the skills and software knowledge preferred. In addition, the collection was limited to twenty offers for Portugal, and about a third of this amount to offers from each one of the other countries covered in the research that was done.

2.1. National context (Portugal)

The analysis in this section addresses the opportunities for integration into working life of industrial designers through an Internet search of job offers during the extended time period from October 2010 to February 2011, in Portugal. Job offers for positions in the field of industrial, product and equipment design were sought and identified. In Portugal, the regions that have a higher rate of development in relation to industrialization, are those where there is a greater number of job offers for industrial designers to be found, in particular, the central region (in and around the cities of Lisbon and Leiria) and the Northern region (in and around the cities of Porto and Braga). Among the most requested features in the offers were product designers with emphasis on the development of innovative products that would make a difference by strengthening the capacity of differentiation in highly competitive environments.

The characteristics and skills most valued are holding at least a degree in design (bachelors and masters are also required in some cases) and having experience in mastering the instrumental and technical means necessary for the conceptual development of new products. Designers must especially master the use of software such as Illustrator, Photoshop™ and 3D tools (Solidworks™, Rhino™, ProE™). Most advertisements aim at designers who have the ability to speak multiple languages, mainly English but also French, German and Spanish, besides their native language.

Regarding interpersonal and relational skills, applicants must show that they are dynamic, balanced, revealing a character that is suitable to work as a team member and having an entrepreneurial, creative and innovative attitude is especially valued. Almost every offer seeks experienced professionals with at least 2 years of experience, to serve in the furniture and electronics industries, among others. The search for trainees (without experience) was not found to be very expressive.

In many job offers, the prospective employer was not specified, and with regard to the benefits (wages) in return for the services to be rendered, it is noted that only a few of the job offers surveyed did declare wage values.

2.2. International context

The analysis carried out in this section discusses the opportunities for integration into working life of industrial designers which were unveiled through a survey of international job offers available on the Internet, in the period from October 2010 to February 2011. The survey focused on job offers in the area of industrial, product and equipment design, identifying the positions advertised, the prerequisites for filling each vacancy and the benefits offered. United States, Australia, Germany, France, Italy, UK, Canada, Spain and Brazil were the countries included, given the perception of the authors that in these countries there are many companies looking for professionals in the field of industrial design. Regarding Portuguese speaking countries and territories, other than Angola, Brazil and Portugal, including Cape Verde, Mozambique, Macau, East Timor, Guinea-Bissau and São Tomé and Príncipe, the demand for industrial designers does not seem to have much expression, judging from the period when job offers placed on the Internet were surveyed.

Among the most requested skills for the industrial designer, emphasis appears in many cases on packaging design and development of innovative products. The most valued qualifications concern holding at least a Bachelor degree in a design specialization. Besides up-to-date skills in 3D solid modelling, some companies place a premium on knowledge of CATIA software and having good overall designer computer skills, including experience with webpage design. Almost every offer surveyed sought experienced professionals with at least 3 years of experience.

2.3. Discussion on opportunities for integration in active life

The survey carried out at national level about opportunities for integration in working life for industrial designers showed that, with regard to experience, companies often want designers with some experience and rarely ask for those looking for their first job. In relation to compensation, very few companies disclosed earning ranges in their offers.

International demand for industrial designers is significant, both in European countries and globally. In what concerns Portuguese-speaking countries, with the exception of Brazil and

Portugal, the demand does not seem to be expressive. Regarding remuneration, almost all companies disclosed the salary offered, with the monthly value changing from country to country and mostly between 1000 and 3000 USD. The most common areas of action ranged from electronics to footwear, including household appliances, among others. In terms of experience, companies often seek designers with some experience and rarely ask for those seeking their first job.

3. Importance of entrepreneurship for the success of the industrial designer

The development of innovative products and improving existing products is a complex business effort, involving the integration of various skills, from design to manufacture through design and marketing. The practice of entrepreneurship appears to be increasing worldwide as a career option, against the backdrop of socio-economic difficulties undermining many economies and countries and reducing opportunities for those wishing to enter the labour market. However, the practice of entrepreneurship coexists with the failure of many organizations, due to the low levels of education of entrepreneurs and heir reduced motivation to use management tools that would enable them to upgrade their activities.

Design has proved an important tool in creating value through innovation. The growing appreciation of the profession in the world has highlighted the urgent need of education in design to be targeted to deal with the globalized and extremely competitive economy and job market. In this sense, design education must be addressed as to support and nurture young designers, facilitating their entry in the job market, along with the dynamism and expansion of design as a factor in creating value for companies and nations.

As an example of the pathways of professional affirmation of designers, entrepreneurship, at various stages of the careers of designers, can be studied from both a theoretical standpoint, and considering a national reality. In this context, and by way of example, this section of the paper presents a technological project for sustainable design and outlines the contours of a proposal for social entrepreneurship based on that project, aiming to contribute to community development.

3.1. Sustainable design project with a view to entrepreneurship in collabouration with the local community (Niassa – Mozambique) to contribute to local development

This section presents a design project with a view to sustainable entrepreneurship, entitled "Development of sustainable integrated solutions for the northern provinces of Mozambique - Ecotourism." The project consisted of the development (analysis and design) of a hot air balloon and an infra- structure for the production of biogas with concerns for environmental sustainability, to support ecotourism in a group of communities. For this project, several objectives were thus outlined, with key points that should be achieved, such as the use of ecological materials, achieving an aesthetically acceptable, innovative, affordable, competitive and uniform design.

Ecotourism is a segment of tourism that explores the environmental, natural and cultural heritage, encourages conservation and seeks the formation of environmental awareness through the interpretation of the environment and promoting the welfare of populations. In this project, it was chosen to develop a hot air balloon. The idea was to create the balloon to support human recreational activity, through geographical exploration from an aerial perspective in a non-polluting way, and thus contribute to rural poverty alleviation and sustainable development in rural communities of the northern provinces of Mozambique, and Niassa in particular. The aim was to create a sightseeing tour promoting touristic housing where Eco tourists would be able to stay with the native population, and partake in the activities of these traditional communities. Through this practice, the income would go directly to families and community funded projects in health, education, road construction, among other domains.

Mozambique is a country in the east coast of southern Africa, bordered to the north by Zambia, Malawi and Tanzania, to the east by the Mozambique Channel and the Indian Ocean, and to the south and to the southwest by South Africa and to the west by Swaziland and Zimbabwe. The province of Niassa was selected for the realization of the ecotourism route conceived (approximately 120 km) - Niassa - Lichinga - Dias - Maniamba - Metangula.

The chosen route begins in Lichinga, the capital of Niassa. The first stop would be Dias, about 46 km away. The next stop would be Maniamba, about 43 km from Dias. Departing from Maniamba, the next destination would be Metangula, about 30 km away. Given that the maximum speed for safe travel by balloon is up to 20 km/h (always depending on the weather and wind) and that the autonomy of the balloon is usually 3 hours of flight, it is expected that the route chosen would enable a comfortable and safe journey through the air.

As an air vessel without mechanical handling and dependent on wind currents, weather monitoring becomes essential for the safe displacement of the hot air balloon. The ideal conditions for balloon flights are early in the day, with great visibility and light winds, with speeds up to 10 knots (18.5 km / h or 5.2 m / s). The wind regime in Mozambique is essentially influenced by the circulation of the atmosphere in southern Africa and the main currents blow from southeast. Mean annual temperatures vary between 23° C and 26° C. In areas of high altitude, these are less than 23° C. The warmest and most moist months are October to February.

An anaerobic digester is a device used to produce biogas, a mixture of gases, mainly methane, produced by bacteria that digest organic matter under anaerobic conditions (in the absence of oxygen). A bio-digester is a chemical reactor in which chemical reactions of biological origin take place. The biogas can be used as fuel for the hot air balloon instead of natural gas or gas obtained from oil, both extracted from mineral reserves.

This project was presented on July 8th, 2010 in Maputo, Mozambique, at the international symposium "Towards a research agenda for development ergonomics in Mozambique". Other projects were developed onwards as a consequence of the afore-mentioned symposium and were reported by Couvinhas et al. (2012) and by Coelho et al. (2012).

The results to be achieved with this Ecotourism project based on a hot air balloon that runs on biogas created from bio-digesters placed locally (fueled by organic waste from livestock) include enabling a scientific, historical and cultural basis for the stay of tourists in rural communities, while practicing a kind of eco-tourism, but one that is also exciting and adventurous. Bio-digesters and tourist facilities are to be placed on ground stations, from village to village, to foster increased income for communities and hence an economic boost. Image renders of the project developed are shown in Figures 1 to 3.

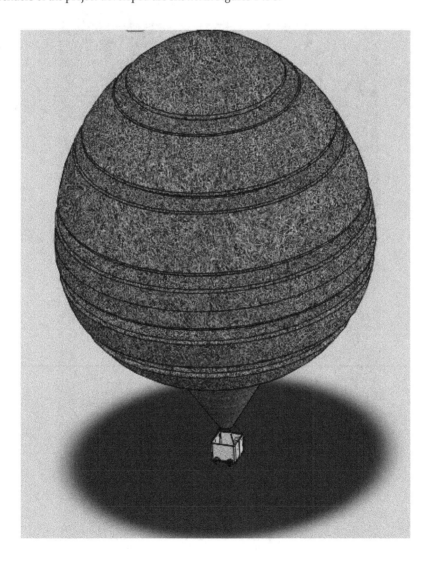

Figure 1. Render of Hot Air Balloon Design

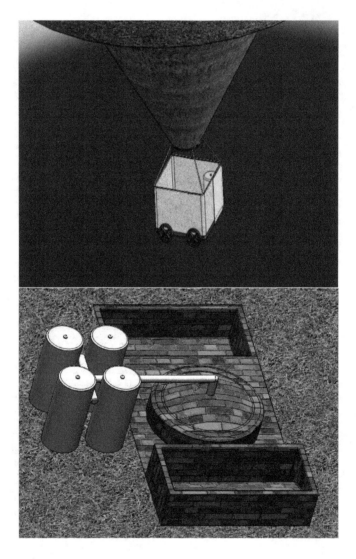

Figure 2. Renders – Basket and Bio-digester

Figure 3. Image *Renders* of Bio-digester, Bicycle and Basket

3.2. Discussion on design and entrepreneurship

The elements gathered and discussed enable considering that in the current context, and the geographic scope considered, entrepreneurship is important for the affirmation of the industrial designer, since it is a way of setting up business structures, albeit small or very small, which allows the development of a corporate culture aligned with the activity of industrial designers. The development of this culture itself can sometimes be difficult in larger and established corporate structures, with other cultures prevailing in these cases (e.g., management, economics, and engineering). This clash of mentalities and forms of action among professionals may in some cases jeopardize productivity and personal fulfillment of industrial designers in the aspects that are paradoxically more valued (creativity and innovation). Thus, entrepreneurship in industrial design can be seen as a way to foster a culture of innovation and creativity, benefiting all stakeholders in the socio-economic fabric, as in many cases these micro and small businesses created by designers serve large companies, while maintaining their independence, or even have the ability to directly generate other profitable businesses based on their innovative projects. This section has presented a project by the first author, a venture into entrepreneurship based on a comprehensive concept for ecotourism and community development.

4. Information collected from surveys

This section presents the condensed results of two surveys conducted. Complete results can be found in Rodrigues (2011). Entry into the labour market occurs relatively early, as most designers surveyed did not wait more than two years to have their first professional experience, which usually occurred in small companies. One of the main difficulties faced by these professionals was working with software that was not known to them. This factor is also considered by employers as one of the biggest problems that designers face when they seek to succeed in the labour market.

In relation to the entrepreneurial spirit of the surveyed designers, the study demonstrated that most of the designers who had had an innovative idea or business during academic training, and when trying to realize this idea had been unable to get financial support for their development, at present, these professionals, come to think about having their own business and thus creating their own jobs. On the analysis of employment held, designer respondents hold a positive spirit, since the majority considered that they had stable employment prospects. As for their salary, most designers think that the salary they receive is not appropriate to the functions they perform.

In terms of the assessment of training for designers, employers positively assess, in general, the training in relation to software taught, though most of them think that more recent software should also be included. Regarding the development of entrepreneurial skills and knowledge of current and emerging technologies, these are aspects that employers consider that need to be reinforced in the training of these professionals.

Considering the answers given by the designers, these suggest that they acknowledge the lack of close contact with business reality during their higher education. The reason for this lack of contact is related to teaching that focuses on many cases in a set of theoretical knowledge that often are not demonstrated or tested by future designers and does not focus enough on practice in the field.

In terms of skills needed to succeed in today's job market and what can be expected for the near future, it is important to highlight some skills related to team performance, namely, the ability to work as a team. Interestingly, throughout the duration of academic training, the ability to work in a team is valued more by teachers than by student-designers, but this capability becomes more valued by designers when they enter working life.

Currently, what is sought in the performance of designers is the mastery of technical skills combined with knowledge about new technologies, but also a focus on creativity and innovation capacity paired with mastery of a wide range of interpersonal skills to enabling adaptation to the evolving context of professional duties.

4.1. Skills most valued per industrial design specialty

For each type of trajectory in industrial design, there exists one or more important and beneficial skill and competency for every type of career path. There are several types of paths, activities or areas of expertise where industrial designers specialize in, such as furniture, footwear, electronics, appliances, toys, jewellery, packaging, automotive, among others.

For the design of furniture, for example, according to data collected in the questionnaire to employers, the skills most valued were knowledge of materials, creativity and CAD, while the audiovisual editing or video editing skills were not very much valued. For the design of footwear, the skills most valued are CAD, while the skills of image editing and art were not highly valued. For the design of electronic equipment, mastery of the area of materials, CAD and prototyping were the skills most valued, while the art and video editing skills were not valued. For the design of appliances, skills most valued are mastery of CAD, materials, prototyping and design methodologies. For jewelry design, skills most valued are talent and creativity. For the design of toys, the skills that are most valued concern the area of materials and creativity. Expertise in packaging design, should be grounded on knowledge of types of materials, on CAD mastery and on creativity.

4.2. Discussion on training and skills top ranked for each type of specialization

Designers and their employers value a variety of factors and skills, with emphasis on creativity and the ability to bring innovation directed to increased market penetration and shares. Designers value the skills that will help them stand out and gain prestige and professional recognition. Designers are required to possess not only technical skills, but also mastery of methodologies for the execution and implementation of design projects and knowledge of the manufacturing methods.

5. Conclusion

The work reported in this chapter shed light on the content of job offers in the area of industrial design, analyzed in relation to skills, qualifications and experience required. Through the surveys conducted, empirical data was gathered that supplemented literature review of studies to enhance the attainment of the objectives proposed for the study. Most designer respondents aspire to become professionally independent of their employer, and to create their own business.

This chapter aimed to analyze and understand the employment opportunities for the profession of industrial designer and their qualifications that are most conducive to recruitment, adopting a national perspective and an international perspective. To achieve this, a two-pronged approach was developed aimed, firstly, to identify pathways to integration in the labour market by industrial designers at the national level, within the Portuguese language space and at the international level, in order to understand the importance of training and other conditioning factors for these paths. On the other hand, work proceeded through surveys to enable the unveiling of the training components seen as critical to employers, and at the same time, this was done to try to envision their future needs in order to provide feedback to the training institutions and enhance effective training and consequently employability. As an example of the pathways of professional affirmation of designers, entrepreneurship, at various stages of the careers of designers, has been studied both from a theoretical standpoint, and considering the socio-economic reality. In this context, and by way of example, the chapter presented a technological project for sustainable design and outlined the contours of a proposal for social entrepreneurship based on that project, aiming to contribute to community development. The work developed and reported in this chapter shed light on a perspective of current job offers in this area, analyzed in relation to the skills, qualifications and experience required. Based on two questionnaire studies made (one questionnaire was addressed to designers active in the Portuguese language space, with 141 respondents, and the other was addressed to heads of enterprises within the Portuguese language space (responses were only collected in Portugal) employing designers, with 19 responses collected), empirical contributions were collected to complement the literature review studies conducted to enhance the attainment of the objectives proposed for the study.

With this work, it was possible to develop an empirical perspective of the challenges facing industrial designers' early career, focusing in particular on designers trained in Portugal. Moreover, the study presents contributions for designers, entrepreneurs, companies and schools that provide training in this area, which may help these groups with tangled interests to pursue their goals more effectively. As civilization makes the transition from the industrial era to one of sustainability, educational leaders around the globe ought to implement a learning system that prepares its young people for life in a unified society. Zinser (2012) provides a starting point to explore what skills and concepts students should be studying to lead the future.

6. Future work

After the presentation of the findings it is important to highlight some issues that provide some recommendations for future research. Based on the results that were achieved, there are clues about what to explore in future research.

As proposals for future work, springing from the results of the present work, one may consider:

- Development of empirical reality-based assessment of the design methodologies used by the designers and companies that employ them.

- Exploration of any differences controlled for gender between objectives and practices found in businesses and especially in companies run by designers.

Moreover, some suggestions for implementation by institutions to improve and adapt to meet the challenges encountered in academic education of industrial designers may be outlined:

- Universities / schools should seek to monitor the changes taking place in the employment marketplace, checking and continuously adjusting their programmes in industrial design to the market, in continuity and in anticipation, in order to respond more effectively to the level of training of designers demanded by employers.

- Training of designers should no longer be exclusively focused on the mastery of techniques, technologies and methodologies, which in itself is no longer enough, to become much more focused on knowledge of how to carry out interactions, e.g. learning to communicate, learning how to lead, learning and practicing teamwork, as well as enhancing innovation and creativity skills.

- Institutions should also focus increasingly on acquisition and mastery of technical skills, maintaining and even increasing incidence on personalized design skills.

- Providing students with intensified and perseverant contacts and experiences with the reality of companies, either through internships or by carrying out real projects that are developed in partnership with businesses, although of an intrinsic curricular nature.

Acknowledgement

The work reported in this chapter was part of the Master of Science Thesis of the first author, supervised by the second author.

Author details

Inalda A. L. L. M. Rodrigues and Denis A. Coelho

Universidade da Beira Interior, Portugal

References

[1] Almeida, João Ferreira de Pinto and José Madureira (1995), Investigação nas Ciências Sociais (Research in the Social Sciences), Editorial Presença, Lisbon.

[2] Carmo, Hermano and Manuela Malheiro Ferreira (1998). Research Methodology: a guide for self-learning, Open University, Lisbon.

[3] Coelho, Denis. A., Patricia, R., Ferrara, Ana. F., & Couvinhas, Tânia. M. Lima and Jake K. Walter (2012). Macroergonomic aspects in the design of development programs in IDCs. WorkA Journal of Prevention, Assessment and Rehabilitation, 41, 2651-2655.

[4] Couvinhas, Ana. F., Patricia, R., Ferrara, Denis. A., Coelho, Sinezia., Jorge, & Jake, K. Walter (2012). Ergonomic considerations for a systemic approach: the millennium maize mills project in northern Mozambique. WorkA Journal of Prevention, Assessment and Rehabilitation, , 41, 568-575.

[5] Rodrigues, Inalda Araci do Livramento Lopes Mota (2011). Prospects for integration into the labour market of Industrial Designers. Dissertation for the Degree of Master in Industrial Design Engineering, Department of Electromechanical Engineering, Faculty of Engineering, University of Beira Interior, Covilhã, Portugal, 128pp.

[6] Zinser, Richard (2012). A curriculum model of a foundation for educating the global citizens of the future. On the Horizon, Iss.: 1, 20, 64-73.

Chapter 10

Design for Automotive Panels Supported by an Expert System

Chun-Fong You, Chin-Ren Jeng and Kun-Yu Liu

Additional information is available at the end of the chapter

1. Introduction

The production process for automotive panels (Fig. 1) has changed dramatically with advances in computer technology. To shorten the automotive development schedule, industry has been using computer-aided design (CAD) and digital model analysis to replace the traditional design method based on human experience. This can reduce the design error rate and improve production efficiency (Choi et al., 1999).

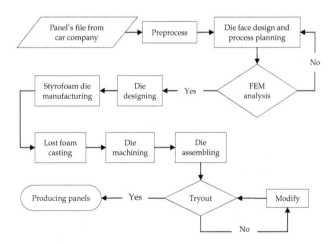

Figure 1. Production process for automotive panels

The entire process from obtaining a panel's file to producing that panel takes about one year, or even much more time. Die design companies that have the shortest schedules and highest quality can occupy a dominant position in the automotive industry.

The stamping die for automotive panels is a cold stamping die; the input is a plane blank; and the output is the panel shape required by a car company. During production, the punch closes the upper mold and lower mold for various tasks. The goal of process planning is to determine how many dies are needed and the content of each die, including stamping direction, tasks, cam type, and other information that is necessary when designing a die.

This study combines practical experience with an expert system, and focuses mainly on preprocess steps and process planning. The system, which is called computer-aided process planning (CAPP) (Marri et al., 1998), is programmed in Java language. The system uses the Spring Solid System developed by the Solid Model Laboratory, National Taiwan University, as the backbone of the CAD system to read the digital surface model, and then output the die layout using Java3D.

2. Expert system

The concept of artificial intelligence (AI) was proposed in the 1980s, and the processing method for computer information is evolving toward that of the human brain. Because many difficulties are associated with the use of AI, an expert system is used to solve problems in particular fields. Generally, it can provide such information as the judgments of experts. Unlike Dyna-vista and CATIA/VAMOS, which are expert systems developed for die design, no software exists that uses an expert system for process planning.

An expert system mainly consists of a reasoning engine, knowledge database, user interface, and developer interface (Fig. 2).

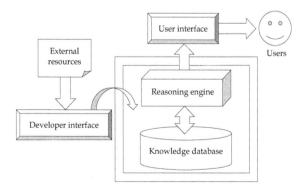

Figure 2. Framework of expert system

1. Knowledge database: This database stores such knowledge as empirical rules, analyzed cases, parameters, and other information used while reasoning.

2. Developer interface: The developer interface allows experts and system developers to modify the knowledge database and reasoning engine from external resources.

3. User interface: This interface allows users to describe questions through a user-friendly operation.

4. Reasoning engine: This engine uses information from the knowledge database to diagnose questions asked by users and search for suitable solutions.

A reasoning engine is widely used with both rule-based reasoning (Lau et al., 2005) and case-based reasoning (Tor et al., 2003; Yuen et al., 2003), and other reasoning methods exist such as neural networks, genetic algorithms, and data mining.

2.1. Rule–based reasoning

The knowledge database of rule-based reasoning stores reasoning rules. After a user enters problems, the reasoning engine starts to reason according to rules and outputs its result (Fig. 3).

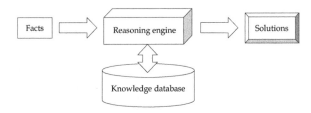

Figure 3. Reasoning process of rule-based reasoning

The judgment rule and boundary rule are typical rules. A judgment rule is represented in the form of "if P then Q," and two types of Boolean and index exist. Judgment by Boolean is used only when two corresponding results exist, and judgment by index is used when more than two results exist (Fig. 4). For instance, a knowledge database contains the rules "if x, then y" and "if y, then z." When a user enters "x is true," the reasoning engine will reason that the result of "z is true."

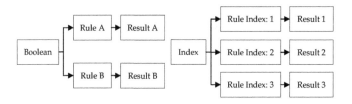

Figure 4. Judgment rule

The boundary rule result is limited by multiple number sets. For instance, if the input number is less than 6.0, 5.0 is output, and if the input number is in the range of 6.0–8.0, 9.0 is output (Fig. 5).

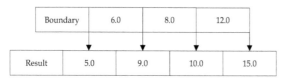

Boundary	6.0	8.0	12.0
Result	5.0	9.0	10.0	15.0

Figure 5. Boundary rule

Using these two rules can increase the number of judgment modes for a system, and enhance its reasoning ability.

The problem of rule-based reasoning is that converting knowledge into rules is difficult. Knowledge can be separated into explicit knowledge and tacit knowledge (Polanyi, 1958). Explicit knowledge can be converted into rules explicitly, while tacit knowledge cannot. The knowledge associated with process planning is almost always tacit knowledge.

2.2. Case–based reasoning

The knowledge database in case-based reasoning stores previously analyzed cases. After users enter a new case, the reasoning engine compares it with all previously analyzed cases in case base, and then searches for the most similar case and reasons for results based on the case (Fig. 6).

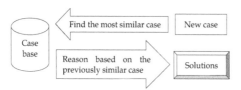

Figure 6. Reasoning process of case-based reasoning

Case-based reasoning has the functions of retrieve, reuse, revise, and retain, called the 4Rs (Kendal & Creen, 2007). After retrieving the most similar case from case base, the information of this case is reused to the new case, and then the proposed solution is revised. Finally, the new case is retained in the case base as a reference for subsequent reasoning (Fig. 7).

3. Die layout design

Automotive panels can be classified into appearance parts and structure parts. Appearance parts can be seen after assembly, including door, hood, fender, roof, and trunk lid; however, structure parts cannot be seen after assembly.

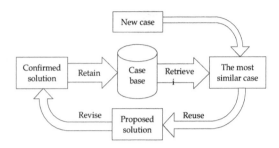

Figure 7. The cycle of case-based reasonin

First, this study uses the left side of a fender (Fig. 8) to illustrate the die layout design process (Fig. 9), including feature recognition, machining center searching, drawing direction optimization, and process planning.

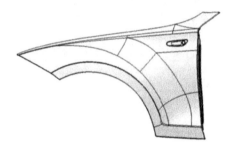

Figure 8. Left side of fender

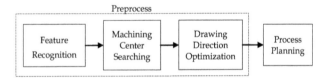

Figure 9. Die layout design process

3.1. Feature recognition

The purpose of feature recognition is to categorize a panel into different sections to establish a bridge between a CAD model and the CAPP system (Zheng et al., 2007) because a panel model without feature recognition is merely unsorted surface data.

If the curvature of single surface exceeds a critical value, it is called a bend surface; otherwise, it is called a flat surface. Bend surface whose curvatures in two domains both exceed critical values is also called a corner surface (Fig. 10).

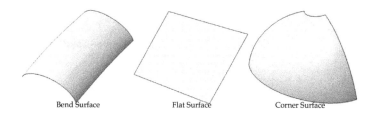

Figure 10. Bend, flat, and corner surface

A group contains parts with the same surface type that are in contact. A panel can be separated into several groups. The group that is shaped during the drawing operation is called the product-in group or main group (Zheng et al., 2007), and the other groups are collectively called the product-out group, which is separated into corners and subgroups (Fig. 11).

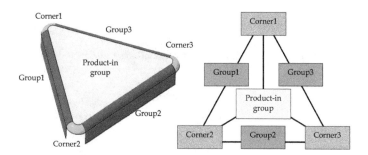

Figure 11. Relationship between features

The flat group with the largest area is a product-in group, and the bend group in contact with the product-in group is the main bend group. Corner surfaces are used to separate the main bend group into several smaller main bend groups, and the flat groups in contact with the main bend groups are the main flat groups. If other groups are in contact with the main flat groups, they are regarded according to the order of bend groups and flat groups (Fig. 12).

The final step in feature recognition is to search for hole features. After finding all edges of a surface, edges shared with another surface are called shared edges; otherwise, they are called single edges (Fig. 13). All single edges for a closed-loop comprise a hole feature; however, the longest closed-loop of single edges is the outer boundary of a panel.

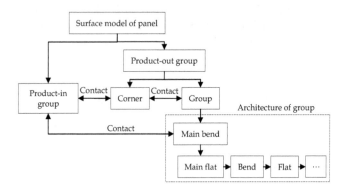

Figure 12. Framework of features

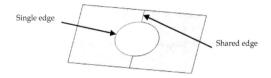

Figure 13. Single edge and shared edge

Finally, the feature recognition result for this sample panel has six groups and six corners (Fig. 14).

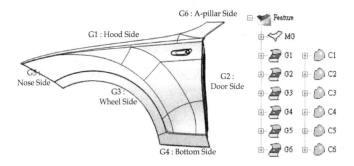

Figure 14. Feature recognition result for fender

3.2. Machining center searching

The file of a panel from a car company is related to the origin of a car (always at the center of the left front wheel, but differs among car companies). Before deciding the drawing direction, one should first search for the machining center as a new origin, which is a reference point of dimensions marked while designing the die and a machining center while assembling the die.

The method of searching for the machining center is to find the minimum bounding box first, and to define the longest side to the shortest side as the x-, y-, and z-axis in sequence. The direction of the x-axis is used as a reference when designing the longest side of a die and the direction of the y-axis is used as a reference when designing the shortest side, which is the feeding direction. Finally, the center of the upper rectangle is regarded as the machining center (Fig. 15).

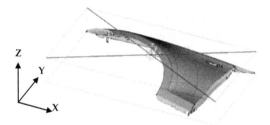

Figure 15. Minimum bounding box and machining center of fender

3.3. Drawing direction optimization

Before introducing the drawing direction optimization method, this study introduces the drawing task. The drawing procedure differs markedly from other tasks. A plane blank is placed on the piston, and the punch drives the upper die downward to clamp the blank with the piston, and then continues its downward movement with the piston to form the blank with the lower die (Fig. 16); notably, other operations are conducted without the piston.

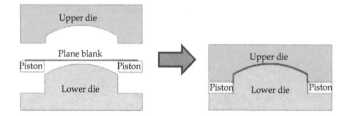

Figure 16. Procedure of drawing operation

The drawing task is always the first operation, and it is the most important task because it shapes the entire product-in group and a little other groups. The key factors in the drawing task are stamping direction and modeling of the die face (You et al., 2011). The stamping direction affects the forming ratio of a panel and product quality, and the modeling of the die face affects the difficulty of follow-up tasks and number of operations. Only after one identifies the drawing direction can the die face be designed and the process planned.

Drawing direction optimization can be summarized using the following three principles: minimum depth, equal angle, and without an undercut. These principles are only for the product-in group, not other groups, because forming the product-out group is not within the scope of the drawing operation.

1.   Minimum depth

Drawing depth is the distance on a panel in the drawing direction (Fig. 17). A large depth can cause cracking and increase the height of a die, thereby increasing cost. Thus, minimizing drawing depth can reduce the degree of cracking; and it means that a shallow drawing is used instead of deep drawing.

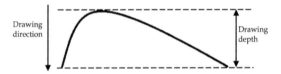

Figure 17. Drawing depth

The method of searching the drawing direction with the minimum depth divides the range 0–180° into five equal parts (i.e., 0°, 45°, 90°, 135°, and 180°), and the depth in each direction is calculated. If the depth in the first direction is the shallowest, then the first direction and second direction are divided into five equal portions and each depth is calculated again. If the depth in the second direction is the shallowest, then the first direction and third direction are divided into five equal portions and each depth is calculated again (Fig. 18); this process continues until the search range converges to <0.1 to determine the angle rotated along the x-axis and y-axis, and the direction of z-axis after rotating is the drawing direction with minimum drawing depth.

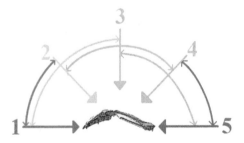

Figure 18. Searching drawing direction with minimum depth

2.   Equal angle

Characteristic lines (Fig. 19) are very important in the product-in group of appearance parts. If characteristic lines are offset from the original position, it will be very obvious from the outside of a vehicle.

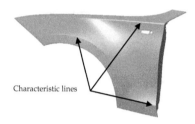

Characteristic lines

Figure 19. Characteristic lines on fender

The reasons for the offset of characteristic lines are that non-uniform forces exist on both sides of characteristic lines. When the slope of one side is larger than that of the other side, characteristic lines will be offset to the more oblique side because the material flow rate is slower than that of the other side. However, automotive panels always have irregular, asymmetric, and complex shapes. The method for preventing offset of characteristic lines is to make all slopes from boundaries to characteristic lines as close as possible (Fig. 20). This study sums all normal vectors in the product-in group and calculates the average normal vector, which is the drawing direction with equal angle.

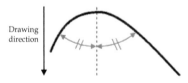

Drawing
direction

Figure 20. Equal angle

3.   Without an undercut

An undercut is an area that cannot be reached by the upper die and lower die during stamping (Fig. 21), and the die will damage at that area. If an undercut area is unavoidable, cams must be used or follow-up operations are needed to shape the undercut area.

In this study, two novel methods are applied for detecting undercut areas after determining the drawing direction, and the range of detecting is not only product-in group but also product-out group because some groups with simple modeling without an undercut are still shaped during the drawing operation.

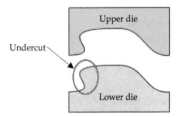

Figure 21. Undercut

The first proposed method calculates the angle between all normal vectors of point data and the drawing direction. If the angle is 0–85°, the area around the point is safe without an undercut. If the angle is 85–90°, and then the area around the point is close to an undercut, such that one should pay special attention to the draft angle. If the angle exceeds 90°, the area around the point is certainly an undercut, such that this area cannot be shaped during this operation, and other tasks are needed to shape this undercut area (Fig. 22).

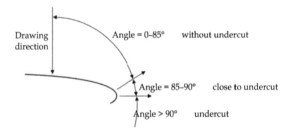

Figure 22. Detecting undercut by angle

This method can determine whether undercut areas exist, but cannot detect the undercut area accurately (Fig. 23). Section (a) is detected correctly as an undercut, but section (b) is not because the angle between the normal vector and drawing direction is <85°. In fact, section (b) still belongs to the undercut area.

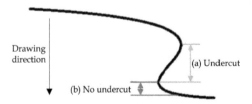

Figure 23. Fail to detect the undercut area by angle

To overcome the detection problem, this study applies another novel method that detects all undercut areas accurately. All point data are adopted as start points and the drawing direction is adopted as the direction vector to establish a ray. If no points exist at the intersection between the ray and the panel, the area around the point is safe. If points exist at this intersection, the area around the point is an undercut (Fig. 24). Section (b) is also detected as an undercut as section (a) correctly by the intersection between the ray and the panel.

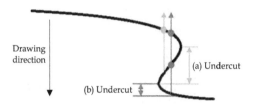

Figure 24. Detect the undercut area correctly by intersection of ray

The drawing direction result for a fender (Fig. 25) is determined from the half minimum depth and half equal angle methods.

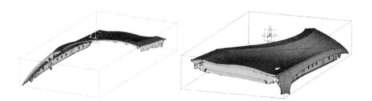

Figure 25. Drawing direction of fender

3.4. Process planning

The purpose of process planning of automotive panels is to identify the number of operations, the tasks in each operation, and the content of each task. This researching proposes an automatic reasoning procedure based on expert experience and the laws of physics. First, the essential tasks based on the feature recognition and drawing direction results are searched and reasoned and then arranged in each operation. Finally, the most suitable stamping direction of each operation is analyzed, and the machining direction of each task is based on the stamping direction.

Common tasks in die layout of automotive panels are drawing, trimming, restriking, flanging, piercing, and burring (Table. 1). These tasks are characterized as follows:

Type	Task	Simplified Description
Forming	Drawing(DR)	Form the product-in group and some other groups
	Flanging(FL)	Flange the unformed groups to position
	Restriking(RST)	Form the groups whose precision is not enough
	Burring(BUR)	Flange the boundary of hole feature
Cutting	Trimming(TR)	Cut the redundant material
	Piercing(PI)	Cut the hole feature

Table 1. Classification of common tasks

1. Trimming task

The trimming task is separated into two parts—one uses a trimming knife to trim outside the surface of a panel, which is called scrap material, and the other trims the scrap material into smaller pieces with the longest diagonal <500 mm to discharge from punch conveniently (Fig. 26). Based on the laws of physics, when the machining direction of the trimming knife is parallel to the normal vector of a trimmed surface, it will apply the optimal trimming force to make the situation of boundary well. If the angle between the machining direction and the normal vector is too large, the boundary will produce deckle edges and sharp phenomenon.

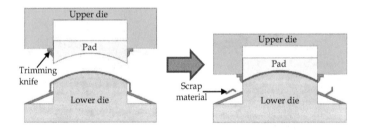

Figure 26. Procedure of trimming task

2. Restriking task

The contact between the restrike knife and surface is face to face (Fig. 27). The restriking task is necessary when surfaces deform because of springback after drawing and trimming, or the accuracy requirement is high because a surface overlaps another surface of another panel during assembly. If the machining direction of the restriking knife is parallel to the normal vector of a surface, the optimal restriking force is applied to the surface.

3. Flanging task

The contact between the flanging knife and surface is a line contact (Fig. 28). The flanging task is necessary when the position between the surface after drawing and the final shape of a panel

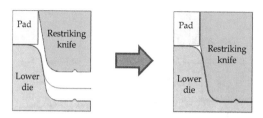

Figure 27. Procedure of restriking task

differ, and there are the effect of restriking when flanging to the end. The machining direction of the flanging knife perpendicular to the normal vector of a surface will apply the optimal flanging force for a surface.

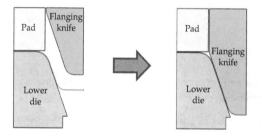

Figure 28. Procedure of flanging task

4.   Piercing task

Each hole feature requires a piercing task (Fig. 29). Hole features can be classified as basic holes, lock holes, and enlarge holes based on their different functions during assembly. A basic hole is for locating the panel, and a lock hole is for locking panels. The accuracy requirement of both holes is high. A enlarged hole whose accuracy requirement is low for passing through the machining tools during assembly. The piercing principle is similar to that of trimming. The machining direction of a drill parallel to the normal vector of a hole feature will also make the situation of boundary of hole well, and the allowed angle is based on the size of the hole feature.

5.   Burring task

The burring task is necessary when bending shapes exist at the boundary of a hole feature after piercing. The burring procedure resembles that of flanging (Fig. 30). The best machining direction is the same as the piercing direction.

This study now introduces the relationships among all tasks and specification of planning.

1.   Trimming is arranged before restriking and flanging

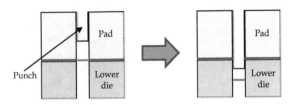

Figure 29. Procedure of piercing task

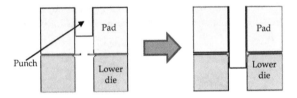

Figure 30. Procedure of burring task

If trimming is performed after restriking and flanging, residual stress will cause severe deformation after trimming.

2. As many trimming tasks as possible are arranged in the same operation

As the number of pieces of scrap metal is typically excessive, they could not be discharged from the punch easily, and the difficulty in die design will increase; however, if trimming tasks are arranged in a backward order, then the restriking and flanging tasks will also be arranged in a backward order, such that the number of operations will increase, thereby increasing production cost.

3. As many piercing tasks as possible are arranged in the same operation

Positional errors always exist in each piercing task. If piercing tasks are conducted in different operations, and then the error among all holes will likely increase because offset directions differ. Thus, arranging piercing tasks in the same operation can reduce error because all offset directions are the same.

4. Piercing is arranged after restriking and flanging

When restriking or flanging, the position, shape, and size of a pierced hole will change; thus, piercing tasks are usually arranged after restriking and flanging. However, enlarge holes whose accuracy is low are acceptable before restriking and flanging to prevent generating an excessive amount of scrap material from holes, which increases discharge difficulty.

5. Cutting tasks are not arranged with upward flanging tasks

If upward flanging tasks exist, the pad is mounted on the lower die. However, it is relatively unstable during production, such that arranging cutting tasks, such as trimming and piercing, will increase the magnitude of errors.

6.   Burring is arranged after piercing

A hole feature with a bending boundary requires two tasks. Although a new task combining piercing and burring exists, it is used rarely as it is associated with increased cost and a short service life; thus, it is not considered by this study.

7.   Determining the machining direction of each task

Using cams increases production cost, such that using the stamping direction as the machining direction is best. Additionally, if difficulty is associated with the stamping direction, cams can be used to change the machining direction. Cams are generally classified as suspension cams and non-suspension cams (Fig. 31). The knife of the former is mounted on the upper die, increasing cost and reducing service life. Thus, this study first considers non-suspension cams. However, if problems in discharging scrap material or feeding the blank exist, then suspension cams are used to increase the space of the lower die.

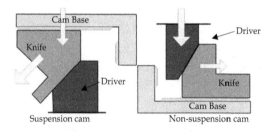

Figure 31. Suspension cams and non-suspension cams

Based on the specifications described above, this study summarizes the sequence of all tasks (Fig. 32). However, the stamping direction and detailed tasks in each operation must still be confirmed according to the panel models.

Figure 32. The sequence among all tasks

Before arranging tasks into operations, one must determine which tasks are needed. Piercing and burring tasks are easier than other tasks to reason. Each hole feature needs a piercing task, and when the normal vector of a hole is not parallel to the normal vector of the boundary, the hole feature also requires a burring task (Fig. 33).

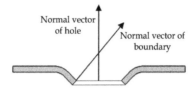

Figure 33. Reasoning whether the hole feature requires a burring task

Hole feature classification is based on the product drawing from a car company. However, this study simplifies this classification to a enlarge hole with a lower accuracy when the boundary length exceeds 50 mm because the area of enlarge holes is much larger than that of other holes.

Because drawing and trimming results affect the assessment of forming tasks, this study arranges drawing and trimming operations first. The first operation is only for the drawing task, such that other tasks are not arranged. In addition to trimming tasks around the panel, this study also considers all piercing tasks of enlarge holes of the product-in group in the second operation, and searches for the stamping direction that maximizes the number of piercing tasks without cams. If any piercing task of an enlarge hole cannot be conducted in the stamping direction, it should be arranged in follow-up operations because cams will interfere with trimming knives.

If some groups cannot be shaped while drawing (e.g., undercuts exist in the drawing direction or the draft angle is too small), this study designs the die face with a shape that can be drawn successfully and the follow-up flanging task is used to shape the groups. Even when groups can be drawn successfully, they cannot be trimmed in the trimming direction because of the normal vector of the trim line, and they still need a flanging task by designing the die face with a shape that can be drawn and trimmed successfully (Fig. 34).

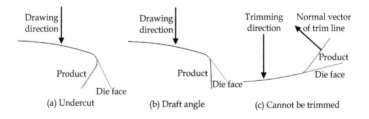

Figure 34. Some reasons for flanging task

If a group can be drawn and trimmed successfully, restriking tasks are only needed to increase accuracy. The complexity of modeling of a group affects the forming result. Applying a flanging task to groups whose modeling is complex will result in cracking or wrinkling easily.

The restriking task result is better than that of flanging task because most modeling is done while drawing.

When a group cannot be drawn or trimmed successfully and modeling is complex, the group is shaped by two forming tasks. A flanging task is first used to bend the die face into a transitional shape and the restriking task is then used to form the transitional shape to product shape (Fig. 35).

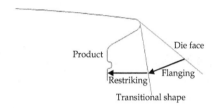

Figure 35. Two forming tasks: restriking after flanging

After drawing and trimming operations, each group requires at least one forming task, such that all remaining piercing tasks are not considered currently, but burring tasks, whose hole features are already pierced, are arranged. The best stamping direction of the forming operation minimizes the number of cams needed and has the best forming effect.

If other second forming tasks exist, they should be arranged after the forming operation. If all forming tasks are arranged, then one must consider the remaining tasks of the hole feature according to the order of piercing and burring tasks, and the best stamping direction for follow-up operations is the same as that in the forming operation.

The sample fender requires five operations in the result of process planning. First two operations are for drawing and trimming, and suitable standard cams are chosen or the size of homemade cams is reasoned based on features (Fig. 36).

The operating time of this example is taken about twenty minutes, and it will change with the file size and panel shape. The actual cost time except die face design from the experience engineers on the factory is taken about three to five days.

4. Sample of structure parts

This section introduces another sample of a structure part (Fig. 37). The procedure and concept of process planning for appearance parts and structure parts differ little. However, the functional differences among structure parts make that drawing direction optimization is needed for the entire panel, including the product-in group and product-out group, rather than for the product-in group only as appearance parts, because there are fewer undercut areas in structure parts and most product-out groups are shaped while drawing. The emphasis for

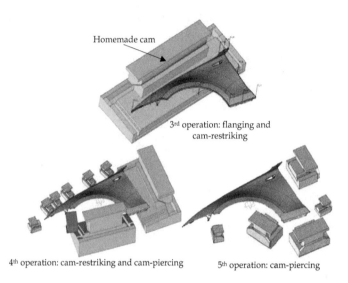

Homemade cam

3rd operation: flanging and
cam-restriking

4th operation: cam-restriking and cam-piercing   5th operation: cam-piercing

Figure 36. Process planning result for fender

Figure 37. Sample of structure parts

structure parts is strength and rigidity, not appearance, such that the offset of characteristic lines is not important. Thus, the equal angle principle is not used to search the drawing direction of structure parts.

Furthermore, feature recognition procedure differs from the appearance parts and structure parts because structure parts do not have the constant modeling rules that appearance parts have. For instance, the area of the product-in group may not be the largest; the main bend group may not constitute a closed-loop; corner surfaces for separating the main bend group into smaller pieces may not exist; and the single surface, even when its curvature exceeds a critical value, should be classified as a flat surface. Thus, the automatic recognition procedure for appearance parts does not apply to structure parts.

Although this system supports manual feature recognition, too much time is needed to click on all surfaces. Thus, this study applies a novel procedure that uses both automatic and manual operations for feature recognition for structure parts. Users must establish only the framework of the features based on panel model and click on one or two surfaces in each group as start surfaces to identify automatically all surfaces with the same feature.

The most significant problem when searching a bend surface is the direction of the connection between bend surfaces (Fig. 38). Although the curvature of surface (a) and (b) both exceed the critical value, surface (a) should be deemed a flat surface instead of a bend surface based on the panel model.

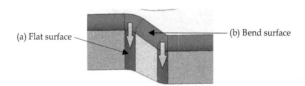

Figure 38. Connecting direction between bend surfaces

First, users click on at least two adjacent surfaces in the bend group. The surface that will be judged to be in the same bend group must satisfy the following three conditions: its curvature exceeds the critical value; it contacts the selected or judged bend surface; and the connecting angle between bend surfaces is less than a critical value (Fig. 39).

Figure 39. Connecting angle between bend surfaces

Fig. 40-42 show feature recognition result, the optimized drawing direction, and process planning results for structure part, respectively.

5. Protocol

The foundations of a knowledge database are knowledge acquisition and knowledge representation. Knowledge engineers retrieve knowledge from experts, books, or other sources, and then represent it on computer systems. The process of retrieving knowledge is called knowledge acquisition, and the process of representing knowledge on a computer system is called knowledge representation.

This research develops feature protocol and process protocol to record feature recognition and process planning results (Lutters et al., 2000), respectively. According to the process level, the

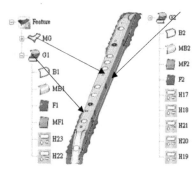

Figure 40. Feature recognition result for structure parts

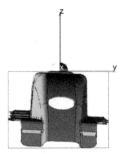

Figure 41. Drawing direction for structure parts

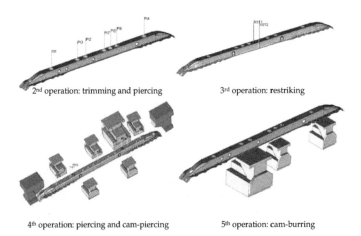

2ⁿᵈ operation: trimming and piercing   3ʳᵈ operation: restriking

4ᵗʰ operation: piercing and cam-piercing   5ᵗʰ operation: cam-burring

Figure 42. Process planning result for structure parts

information is split by characters "! ; : <>,". Each operation should contain an ID, type, stroke, die center, die dimensions, shoe height, and arranged tasks (Fig. 43). Notably, each task type has its own information that must be recorded (Fig. 44).

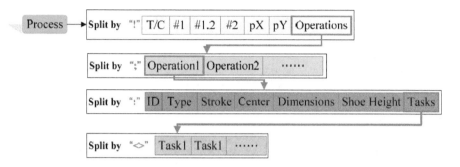

Figure 43. Process protocol

Figure 44. Protocol of each task

6. Conclusion

The process planning result is not unique, and such results will vary based on specifications from different die design companies and different designers. If a die design company and its designers are the same, even when the panel is the same, process planning results will also vary due to different client demands and different cost considerations.

Therefore, this study is based on the premise of compliance with the principles set forth in section 3.4 to reason the acceptable and enforceable process, and allow users to modify the content of process manually in response to different conditions and circumstances (Chapman & Pinfold, 1999; Ciurana et al., 2006).

Acknowledgements

The authors would like to thank the engineers of Geo Ho International Co., Ltd. for their technical supports. Ted Knoy is appreciated for his editorial assistance. The authors would like to thank the National Science Council of the Republic of China, Taiwan, for financially supporting this research under Contract No. NSC-98-2221-E-002-135-MY2.

Author details

Chun-Fong You, Chin-Ren Jeng and Kun-Yu Liu

Department of Mechanical Engineering, National Taiwan University, Taiwan, ROC

References

[1] Ammar-Khodja, S.; Perry, N. & Bernard, A. (2008). Processing knowledge to support knowledge-based engineering systems specification, *Concurrent Engineering*, Vol.16, No.1, pp. 89-101

[2] Chapman, C.B. & Pinfold M. (1999). Design engineering – a need to rethink the solution using knowledge based engineering, *Knowledge-Based Systems*, Vol.12, Issues 5-6, pp. 239-245

[3] Choi, J.C.; Kim, B.M. & Kim, C. (1999). An automated progressive process planning and die design and working system for blanking or piercing and bending of a sheet metal product, *International Journal of Advanced Manufacturing Technology*, Vol.15, Issue 7, pp. 485-497

[4] Ciurana, J.; Ferrer, I. & Gao, J.X. (2006). Activity model and computer aided system for defining sheet metal process planning, *Journal of Materials Processing Technology*, Vol. 173, Issue 2, pp. 213-222

[5] Kendel, S.L. & Creen, M. (2007). *An introduction to knowledge engineering*, Springer-Verlag London Limited, ISBN 1846284759

[6] Lau, H.C.W.; Lee, C.K.M.; Jiang, B.; Hui, I.K. & Pun, K.F. (2005). Development of a computer-integrated system to support CAD or CAPP, *International Journal of Advanced Manufacturing Technology*, Vol.26, Issues 9-10, pp. 1032-1042

[7] Lutters, D.; Brinke, E. ten; Streppel, A.H. & Kals, H.J.J. (2000). Computer aided process planning for sheet metal based on information management, *Journal of Materials Processing Technology*, Issue 1, pp. 120-127

[8] Marri, H.B.; Gunasekaran, A. & Grieve, R.J. (1998). Computer-aided process planning: A state of art, *International Journal of Advanced Manufacturing Technology*, Vol.14, Issue 4, pp. 261-268

[9] Polanyi, M. (1958). *Personal knowledge: toward a post-critical philosophy*, The University of Chicago Press, Chicago

[10] Radhakrishnan, R.; Amsalu, A.; Kamran, M. & Nnaji, B.O. (1996). Design rule checker for sheet metal components using medial axis transformation and geometric reasoning, *Journal of Manufacturing Systems*, Vol.15, Issue 3, pp. 179-189

[11] Tor, S.B.; Britton, G.A. & Zhang, W.Y. (2003). Indexing and retrieval in metal stamping die design using case-based reasoning, *Journal of Computing and Information Science in Engineering*, Vol.3, Issue 4, pp. 353-362

[12] You, C.F.; Yang, Y.H. & Wang, D.K. (2011). Knowledge-Based Engineering Supporting Die Face Design of Automotive Panels, *Industrial Design - New Frontiers*, pp. 21-38, ISBN 978-953-307-622-5

[13] Yuen, C.F.; Wong, S.Y. & Venuvinod, Patri K. (2003). Development of a generic computer-aided process planning support system, *Journal of Materials Processing Technology*, Vol.139, Issues 1-3, pp. 394-401

[14] Zheng, J.; Wang, Y. & Li, Z. (2007). KBE-based stamping process paths generated for automobile panels, *International Journal of Advanced Manufacturing Technology*, Vol.31, No.7-8, pp. 663-672, ISSN 0268-3768.

Chapter 11

Visual and Material Culture in the Context of Industrial Design: The Contemporary Nigerian Experience

I.B. Kashim

Additional information is available at the end of the chapter

1. Introduction

Industrial design is viewed as a synergy between applied art and science aiming at creating and developing aesthetic, ergonomic and functional values in produced artefacts. In the evolution of visual designs in Nigeria, a craft-based design practice has been most prominently featured with the culture acting as a motivating factor. The craft designs in traditional Nigerian settings consist of three dimensional elements such as the object form and the two dimensional features such as patterns, lines and colours. The visual designs on material artefacts have consistently infused culture-oriented aesthetics, thus adding to their local identity and commercial value for increase marketability. For the purpose of this paper, visual and materials culture is viewed as a direct application of industrial design with relevance to the productions of indigenous artefacts which are accomplished with technical expertise and covering wide areas of applied art and design, such as jewellery, interior design, ceramics, household wares, architecture, textile designs, leisure goods and woodwork. The Nigerian concept of industrial design embraces the creation of functional designs with intrinsic aesthetic satisfaction. This article explores the expanding field of material and visual culture insofar as it contribute to events, situations or other features relevant to defining human landscape as well as the social, cultural and natural environments, objects, images, ideas and practices.

In general, material culture can be defined as whole objects that are close to art in characteristics but which do not fully assume art's unadulterated-status. The study of material goods and artefacts, technology and other aspects of material culture have been given systematic attention, especially with prototype invention incorporated into the more generalized fields of work organizations, informal settings, cultural production and domestic settings, etc. All of these cause practical utility and aesthetic value to intersect in influencing material goods and the demand of conspicuous consumption.

Crafts are symbols of Nigerian material culture and spiritual heritage. They are integrated into the living pattern of Nigerians as prized objects for the promotion and preservation of its tourism industry. Conservative estimates indicate that over 70% of the total population of Africa are rural dwellers, engaged in farming along with various craft production activities including metal craft, pottery, carving, textile, weaving, embroidery, leather work, calabash decoration, blacksmithing, bronze and brass casting and tie and dye. These constitute the major rural industries in the form of small- and medium-scale enterprises which are fundamental to stimulating the economic and industrial development of products for mass consumption. According to Ogunduyile and Akinbogun (2006), the introduction of industrial design into the school programme in Nigeria focused on promoting the developing small- and medium-scale enterprises that are expected to compete favourably with imported products, thereby, opening an avenue for creativity, innovation, jobs and wealth creation.

A comprehensive overview of the aforementioned areas has prompted various investigations - using the factors mentioned below - on which the contemporary status of industrial design programmes in tertiary institutions in Nigeria are based upon:

1. The development of prototypes and product systems that can be applied in satellite industries.

2. The application of modelling and simulation for the rapid prototyping of design products.

3. The fast-tracking of industrial growth and the transformation of raw materials into useful products.

4. The provision of appropriate local machinery.

2. Conceptual framework of the study

Africa is highly affected by the creative economic industries. The visual and material culture is influenced by the creative industries which in turn play a critical role in potentially contributing to Nigeria's economy. These contributions are significant to the issues of the eradication of poverty, rural and urban development, community livelihood and survival strategies. The research questions adopted survey tools to investigate and establish the relationship between material culture, artefacts, the creative industries and the commercial values of products. Interviews were conducted to establish the commercial value of the downstream activities of applied art and the crafts in their selection for their commercial value. Apart from interview questionnaires, schedules were used to collect in-depth information and data using structured, semi-structured and unstructured questionnaires. They were designed for a census of creative activities using sampling techniques.

3. Research questions

For the purpose of this study, the following critical questions arose, namely:

a. What are Nigerian creative industries and products about?

b. What are the relationships between art, culture and crafts?

c. What are the threats and solutions to the mass production of traditional crafts?

d. Are the contributions of art and crafts industry significant to the Nigerian economy?

e. What is the interventional role of government in the cottage industries?

4. Research methodology

The research questions were adopted to test the research aim. Quantitative, qualitative and experimental approaches were used as tool of investigation in order to establish the relationship between material culture, artefacts, the creative industries and the commercial values of products. Interviews were conducted to establish the commercial value of the downstream activities of the applied artists and craftsmen in their selection for their commercial value. Apart from interviews, scheduled questionnaires were administered so as to collect in-depth information and data. They take the form of structured, semi-structured and unstructured questionnaires designed for a census of creative activities using sampling techniques.

5. Conceptual definitions of visual and material culture

Culture is that which defines the way of life of a group of people and their interactions with the environment over a period of time. Moving from the abstract to the concrete and from the material realm to the immaterial domain, culture could be described as a thread that holds what a society finds valuable, meaningful and appreciable. Following the models of previous studies (Stephan, 2004; Schein, 1999; Lee, 2004; Hampden-Turner and Trompenaars, 1997; Spencer-Oatey, 2000 in Moalosi, Popovic & Hickling-Hudson, 2007), culture is firmly observed as being dynamic and multi-layered. From the perspective of the intangible elements, Lam et al. (2006) described culture as a set of values (conscious and unconscious) evolved by a group of people living in a society so as to shape that society with specific characteristics, identities, attitudes and behaviours. However, within the tangible layer, culture could also be understood in the social context of artefacts used within a particular environment.

Artefacts, as made-made objects, are a material medium for the communication of cultural values. It includes objects, processes, services and their systems. Since visual and material objects are part and parcel of such communication which gives rise to social forms, visual and material culture has emerged from the interaction between man and artefacts. Today, design artefacts have become an inseparable component of human society, a totem of cultural identity and an important source of reference for modern society. These artefacts are instrumental to aesthetic expression and socio-cultural interaction within a local context.

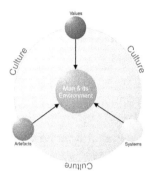

Figure 1. Design connecting the tangible elements of culture.

Behind every man-made object is a valuable concept, beliefs, customs, rituals, habits and ideas that shape its perception within a socio-cultural environment. Visual culture has been described as the artistic tastes and interpretation of what is aesthetically pleasing to an individual depending upon the surroundings and the environment. However, it is not just artistic differences that are influenced by environmental differences. In addition, there is the practicality of the visual statement. Based on the separate goals and purposes that need to be met by aesthetics and the way in which they are interpreted within a specific locale, visual rhetoric widely varies based on classical cultural differences. There are innate differences in the way people view images because our interpretations of aesthetics and practicality stem directly from our localized cultural experiences. The interpretation of design as visual and material culture enables a broad understanding and critical awareness of its meanings as material objects, images and cultural practices that position human beings in time and space.

The term "material culture" is often used by archaeologists as a non-specific way of referring to artefacts or other concrete things left by past cultures. Material culture means something else to art historians. Interestingly enough, when they speak of material culture they look at an object within its environmental and cultural context as things left by past cultures. The organization of cultural activities such as business or economic activities led to cultural industries. The idea of 'creative industries' originated from a concern with the cultural industries. The core of the cultural industry is its creativity (Kwanashee, Aremu, Okoi & Oladokun, 2009).

6. Nigerian visual and material culture

Nigeria is a large society that has different subcultures because it is made up of people bearing different value systems which influence each other to shape a specific culture for the society as a whole. The conventions and values that underlie Nigerian visual and material culture are rooted in the visual art and design forms and the key continuities and changes that characterize development from traditional to contemporary practices.

Industrial design education in Nigeria has focused on culture as a key drive for the design and exploration of materials in the areas of ceramics, textiles and graphical communication in order to express in handcrafted product prototypes both concept and simple design solutions. Industrial design practice tends to promote cultural significance in design formations with reference to local material explorations, ingenious material expressions and the embodiment of cultural values. Industrial design in Nigeria is craft-based and situated in the field of applied art.

7. Contemporary trends in visual and material culture

7.1. Textiles

The local textile industry is diverse, with such branches as spinning, weaving, knitting, sewing, dyeing, embroidery and printing, all of which are significant for clothing, a material culture that is a part and parcel of human life. Common woven fabrics in Nigeria are "Etu", "Alari", "Sanyan", "Fuu" and "Waka", which are used for different purposes ranging from day-to-day and ceremonial use, as in for religious purposes. Cloth weaving is an outstanding craft among the variety of textiles produced in Nigeria. The woven cloths are done on narrow looms in the south-western towns of Iseyin, Oyo, Ogbomoso and Okene in Kogi State; Akwete cloth is made in Akete in Abia State. They are woven materials produced in attractive designs with rich colours and made for male and females. Resistance tie and dye textile products are traditional indigenous crafts done in different societies in Nigeria. Examples are the Hausa traditional indigo dye carried out in the Northern Nigerian areas of Kano, Sokoto and Zaria. The Kano indigo-vegetable dyeing pits are one of the most fascinating aspects of the old city. Various designs are folded into the material before dyeing and the fabric is often beaten to achieve its shiny, attractive appearance. The techniques employed to obtain this look are unmatched around the world. Moreover, although the methods adopted are ancient, these lush works of art on fabric always remain extremely popular and continue to be in great demand.

Figure 2. A typical dye pit at Kano (2012).

Other styles of indigenous textiles are found in the Yoruba towns of Oshogbo, Ilorin, Ondo, Ibadan, Ogbomoso, Oyo and Abeokuta (Eicher, 1976). Local cassava starch is used as a basic resistance material. The blue dye is obtained from a local shrub and dyeing is carried out in large pots at Abeokuta, Ibadan, Ife and Osogbo, while in the northern part of the country it is done in open dye pits which are two to three metres deep. A local vegetable from which an indigo colour is extracted are employed as the colouring agent. This textile craft has been encouraged through the training of the less privileged in such national programmes as Better Life for Rural Women, which was later taken over by the Family Economic and Advancement Programme, set up in 1990. These programmes were meant to develop and encourage the talent and creative energy of women at the grassroots level of society. The Textile Traditional and Research Centre is located at Gbongan / Ife road, Oshogbo, Osun State of Nigeria. The National Research Centre is focused on the textile traditions of Nigeria with a view to encouraging the industry to grow, thereby promoting job opportunities and transforming the rural economy. Below is a sample of the popular *Alari*, as represented by the field research of Maiwada, Dutsenwai and Waziri 2012 (see Fig. 3). The *Alari* that is also *either* referred to as *Aso Oke* or *Aso Ebi*, that is (Commemoration Cloth) was described by Makinde and Ajiboye (2009) as significant attire used for social functions in Yoruba land.

Figure 3. *Agbada (Alari)* produced in Ogun State, Nigeria (source: http//www.aijcrnet.com).

Some other very interesting textile crafts made in Nigeria, especially at Nike gallery, Oshogbo in Osun State, have to do with experimentation with many yarns, which simultaneously change the way the yarn colours are viewed. They are made with a high percentage of wool and 20% or more fancy or strengthening yarns, such as silks, ribbons, cotton or even metallic yarns. Rugs are also made using 100% pure single-type yarns to get warm or beautifully cool cottons. They are designed to fit traditional and contemporary spaces. They are used for floor carpets, picnic rugs and therapy mats, blankets, throws and even duvets, wall hangings and corporate art pieces. Figure 4 shows her gallery logo, her elegant portrait in an elaborate costume and a sample of her batik cloth.

Figure 4. Artist and Designer Nike Davies Okundaye (source: http://www.nikeart.com/main.html).

Ojo (2004) reportedly worked extensively on appliqué and quilting. He described it as a product of the expression of traditional resources which provides an avenue for problem solving in a particular stylistic artistic vacuity and has been identified as bridgehead craft for matching and contacting the confluence of handcraft and the brain. Quilting is about the joining together of layers of fabrics by the tiny running of stitches, while padding is secured between the two outer layers; appliqué is the principle of stitching a multi-colour piece of fabric to a contrasting background in order to illustrate stories depicting communal events.

In culture and musical costumes, they are both used in embellishing masquerades and as garments for creating identity by some Nigerian musicians, such as "Lágbájá". They are used in bedding, cushions, pillowcases and head rests, etc. They are found to match manufacturer's specifications in the mass production industries as featured in areas such as the lapels of footballers' boots, goal keepers' chest and knee protectors, while life jackets, sports wares and hand gloves are now being manufactured using the a textile process that assist in the process of machine quitting technique. A typical example of a culture costume by Lágbájá is shown in figure 6, below. If you were to call out the name Lágbájá in Yoruba, you could be calling for somebody, anybody, everybody or nobody! The Nigerian musician Lágbájá has made a commitment to always wear a mask on stage so as to hide his identity. He is described by the phrase: 'The man without a face who speaks for the people without a voice'. His masks and costumes have many different designs. Generally, Lágbájá's music blends jazz, afrobeat, highlife, juju, funk and traditional Yoruba music, using horns, guitars and keyboards along with traditional Nigerian instruments. Sometimes, the music is purely instrumental but when there are lyrics, they are in Yoruba, English or a combination of the two commonly spoken in Lagos. The lyrics focus on issues relating to democracy and fairness in society; the titles of the songs themselves get the desired message across. (See Fig. 6)

Figure 5. Works Produced at the Traditional Textile Research Centre located at Gbongan / Ife road, Oshogbo, Osun State.

Figure 6. Lágbájá - one of Nigeria's contemporary musicians.

Lágbájá is definitely one of Africa's most exciting contemporary artists, whose elaborate masks and stage costumes link him to the ancient tradition of Egungun: ancestral masquerading spirits who help guide people towards truth and peace.

7.2. Jewellery

Jewellery is very significant to adornment within almost all traditional cultures in Nigeria. Various investigations have been made into beadwork and clothing and other metallic materials used for adornment. According to Adesanya (2010), the jewellery in Yoruba land is made out of different materials, ranging from beads, cowries, plant seeds, annual bows, ivory, leather, stone and metals (including bronze and silver), the latter of which this article places emphasis on as its practice cuts across some major ethnic groups in Nigeria who mass produce them for economic survival. Jewellery and metal work as professions have long historical standing, symbolizing wealth and power. The Nok, Ife, Benin and Igbo-Ukwu have made significant contributions to the development of jewellery. Emeriewen (2007), in his assertion of the paradigm of the Benin Art School experience, refers to the products of metal design as the fabrication of aesthetically functional objects referred to as metal work and craft. He analysed the contribution of the Benin Art School in metal product design in the area of making decorative gongs, lamp stands, flower pot stands, maces and gavels, as well as the contemporary use of metal design for public place furniture. A sample of such metal products in bronze from Benin is highlighted in figure 7, below:

Figure 7. Benin ivory mask (a symbol of Nigerian craft artistry).

7.3. Ceramics

Ibude (2010) has made an extensive exploration into the development of a prototype solid fuel kiln for cost-effective ceramic firing at a Nigerian university. He conceptualized this kiln's production as a critical intervention into the exceptional functioning of the ceramic industries, largely due to the high cost of fuel and the inadequate supply of electricity to meet up with local consumption in Nigeria. The construction of a wood-fired kiln from the researchers' pilot study indicated that it is cheaper to operate in an environment that has significant forest areas and sawmilling activities. He collected the kaolin that he used locally to fabricate a refractory insulator and the dense bricks that he used to build his kiln. The kiln utilizes hardwoods as fuel as it produces better heat (more calories) and the cost of operating a wood kiln was very cheap when compare with gas, electricity and kerosene. The kiln was a prototype that became a source of reference for the development of wood kilns in tertiary institutions in Nigeria and technical manuals.

Adelabu (2011) developed computer-aided ceramic glaze formulation using locally sourced ceramic materials in Nigeria. He focused his study on developing standardized indigenous glaze recipes based on local raw materials in the states of Ondo, Osun and Edo through the aid of a sampled computer-aided software packages namely Matrix V6.01 and Hyperglaze software (authored by Lawrence Ewing and Richard Burkett respectively). This was done so as to establish a better technique for the process of glaze formulation in Nigeria and proffer solutions to prevailing problems of glaze composition encountered by ceramic students and practioners in Nigeria. All of these packages used to develop the Nigerian version of glaze preparation have been developed internationally in order to assist students in their glaze chemistry since the late 1970s. He developed a prototype test kiln that was used to fire the glaze product as highlighted below in figure 8: the product design established awareness as to the current software solution for ceramic glaze experimentation in Nigeria which has not hitherto been used by practicing ceramists in Nigeria.

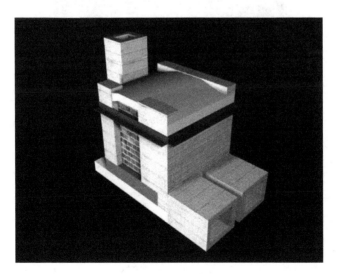

Figure 8. A typical model of a kiln design constructed for the purpose of ceramic firing research (Adelabu O.S, 2011).

Kashim (2004) explored the richly endowed national ceramic resources to develop her capacity for high-tech, value added manufacturing activities in the production of ceramic hardware, such as ball mill lining, porcelain milling balls and pestles and mortars. This production was made through the identification, selection and synthesis of local raw materials available in Nigeria to ascertain their suitability for those whiteware ceramic bodies popularly referred to as porcelain. The prototypes were replicated using the Jigger Jolley machine and the process of slip casting using the plaster of Paris mould. Examples of the works produced include cylinders, pestles and mortars, decorative pieces, ball mill pebbles and lining. An example is the Benin Mask in figure 9, below, which indicates the outcome of the research effort. Levi O'bem Yakubu, a 1979 graduate of industrial design with a focus on ceramics from the famous Ahmadu Bello University (ABU), Zaria, is the founder and chief executive of the Makurdi-based Dajo Pottery Limited, a multi-award winning ceramic industry in Nigeria and world acclaimed company. He has contributed significantly to ceramics production in Nigeria using various production methodologies (see Fig. 10) Other prototype contributions are also in respect of oil burner used to fire ceramic kiln at Federal polytechnic,Auchi. See Fig. 11

Figure 9. A typical cast piece of a Benin Mask in porcelain.

Figure 10. Levi O'bem Yakubu, Chief Executive of Dajo Pottery Limited, Makurdi, Benue State, Nigeria (2012).

Figure 11. Ceramic test kiln, constructed by a Higher National Diploma student of the Federal Polytechnic, Auchi, Edo State, Nigeria (Ogunduyile, S.R, 2006).

7.4. Graphics

Graphical design is all about us, in our daily newspapers, on our commute to work, on book covers and in logos, websites, advertisements, bill boards, product packaging and posters, etc.

Graphics has been used as a tool of communication, like journalists in media organizations, radios and newspapers, and so is an artist in advertising. The mode of communication requires the use of graphical signs and symbols as a medium through which creativity is expressed, generating a societal response that is either positive or negative. Apart from the fact that products are advertised over and over until the public develop a strong feeling of acceptability for them, traditional graphics' inscriptions are influential media products that are vital to society for information dissemination, education, entertaining the public and contributing to the economic development of commercial products. Figure 12a presents a graphical logo and slogan from March 2009 by a former Nigerian President, Umaru Yar'Adua, unveiled as a part of the administration's efforts to rebrand the country's image globally.

Figure 12. A graphical logo for a Nigerian rebranding project.

8. Craft designs

8.1. Calabash decoration

Calabashes - or gourds - are the fruits of several varieties of creeper, some of which are grown along with farm crops. Their creation involve carving and sawing, burning (pyro-engraving) and scorching with heated metal tools, colouring with karan dafi dye and whitening with clay.

Figure 13. Calabash decorations by Cynthia Oldenkamp (source: http://www.uni.edu/gai/Nigeria/Lessons/Calabash.html).

8.2. Woodworks

Osogbo, Benin, Oyo and Akwa have been acknowledged as centres of woodcarving and technology. The carvers have flourished extensively in the southern part of Nigeria from time immemorial, making figures for shrines, masks, portraitures representative of the spirits of the sky, sea, earth, forest, stream, fire and thunder. Many of the old carvers' works are found in museums and public places. Prominent among these woodcarvers is Lamidi Olonade Fakeye (figure 14), who introduces decorative doors into modern architecture (carved doors in low relief) using Yoruba graphic symbology through the mix of traditional interlaces, circles and linear designs based on curves and squares, blending them with figurative images. He was an apprentice to the master carver George Bamidele Arowoogun. In 1978, he became an instructor at the Obafemi Awolowo University in Ile-Ife, Nigeria, where he unveiled his incredible statue of *Odudua* (a Yoruba legend) nine years later. Between 1989 and 1995, Lamidi served as artist-in-residence at several prestigious American universities. He uses a lot of political themes in order to promote national unity among the Yoruba, Hausa and Igbo. Examples of his work can be found at:

1. The Kennedy Centre panel door for the cultural centre, Washington DC, USA.

2. House posts for the Edena gate house of the Oni of Ife's palace, Aderemi Adesoji's at Ile Ife in 1955.

3. Catholic Chapel door at the University of Ibadan.

Figure 14. Great Master Carver Lamidi Fakeye (source: http://37thstate.tumblr.com/post/20410112677/great-master-carver-lamidi-fakeye).

8.3. Cane works

Cane work is one of the major craft industries in Nigeria, with its raw materials of cane and willow sourced from the forest. The production method is conducted through weaving by crossing them over and under one another at right angles in three-dimensional forms, such as for upholstered baby cots, side tables and hamper baskets. The prominent production centres are Lagos, Abeokuta, Ibadan, Onitsha and other commercial centres. The industry possesses the functional capability to serve human needs in homes, offices and outdoor engagements as their processes have unique characteristics which distinguish them from machine-fabricated substitutes (Kayode, 2004).

Figure 15. a) Typical Nigerian cane furniture (2012), b) A Yoruba talking drum with the stick hanged on it, c) Yoruba drummer beating the drum, Source: http://yorupedia.com/subjects/education/yoruba-music/

8.4. Leather works

This is used as a Yoruba talking drum called "Dundun" - a double membrane hourglass-shaped tension drum. They are used in activities such as ritual performances, ceremonies (burials, marriages and chieftaincy), for communication in heralding a visitor's arrival and as a source of information for kings and villagers. The production relied upon wooden frames and leather membrane fastening with rope. They are operated with a drum stick referred to as "Opa Ilu" made from a wood called "Ita" or "Agboyin", which has the characteristics of been curved at one end permanently at such convenient that it can be used to beat the talking drum. The talking drum has a percussion stick (a 1 – ½ foot long beating stick with a curve end knob)

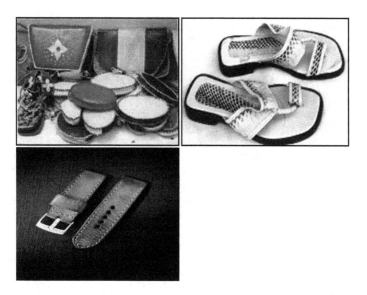

Figure 16. Leather works made in Northern Nigeria (bags, sandals and wristwatch straps).

9. Entrepreneurial product development

The concept of sustainable product development should satisfy public demand for services. It addresses both the demand and supply side of the economic equation. Product designers are traditionally concerned with the relationship between manufactured artefacts and people, enhancing the link between the environment and society through environmentally friendly products. It seeks to do this by attaching sustainable criteria to the quality and value of products.

Skilled and talents are used creatively in the production of traditional crafts items which gave birth to the creative product industries. The industries have their origin in individual creativity, skill and talent, which have a potential for wealth and job creation as well as the generation and exploration of intellectual property (e.g., blacksmithing, goldsmithing, pottery, leatherwork, woodwork and calabash engraving).

The crafts belong to cultural industries which place emphasis on those industries whose inspiration is derived from heritage, traditional knowledge and the artistic element of creativity.

10. The dilemma of the interface between local crafts and modern technology

In most countries with advanced design practices, industrial designs are registered in order to be protected under industrial design law. As a general rule, for registration, the design must be "new" or original. Generally, a rule may vary across countries but "new" means that no other identical or very similar design is known. In the Nigerian context, industrial designs are yet to fully evolve into modern creative expression, despite a vast heritage of a rich visual and material culture which is manifested well in the arts and crafts. With a seeming limitation in access to modern technological design tools, industrial design practice had been side-tracked into placing more emphasis on the profusion of cultural elements in local objects' functions over the development of technical features which are also indispensable tools in optimizing product quality and efficiency. In the recent trend where Nigerian product users are increasingly exposed to the aesthetic and functional capabilities offered by modern design products, the appreciation of local products tends to be disfavoured. This is not to say that local design products are performing poorly on the functional level, but there seems to a technological gap that must be filled by local products in order to favourably compete with the highly standardized design products. Of course, industrial design in Nigeria has the potential to increase the value of manufactured products, improve living conditions and provide the continent with a weapon to wage a war on substandard products from other continents. However, industrial designers must wake up if they are to harness the rich cultural heritage within their borders with the necessary technical know-how so as to become relevant in the socioeconomic advancement of the 21st century (Ogunduyile & Akinbogun, 2006).

11. Conclusion

The concept of sustainable product development in Nigeria captured the subject through notions of development, poverty, trade, population, social and cultural conditions. All of these matter in addressing the supply and demand of economics. This study further observed that early industrial designers' work was primarily focused on physical products but, today, this requires applied behavioural science, rapid prototyping, statistical knowledge and the ability to develop experimental designs.

The study reveals further that the development and diffusion of domestic and improved technologies in cultural products' design within small-scale industries offers a lot in terms of enterprise productivity, employment generation and import substitution. The development of product designs is significant to the economic life of any nation. It has the tendency to increase the value of manufactured products, improve the living conditions of the people and create an appropriate competitive standard against the dumping of foreign goods in Nigeria.

It is therefore recommended that encouraging the commercialization of indigenous craft-based technologies require the following:

1. That a development bank should assist in the funding of projects while commercial banks should provide credit services to cater for entrepreneur with limited resources.

2. The government should encourage the flow of technology-related information, especially for new products, process development, competitive initiatives through the encouragement of tourism (as in Kenya) and access to international market opportunities.

Nigeria is endowed with a diverse and rich visual and material culture which forms the core of its national heritage and which cannot be underestimated in the global context. In the post-colonial era, the fast-growing pace in global design and the technological shift calls for a cultural reinvention and adaptation suitable for contemporary tastes and standards. Local designs will not perform on the global front where they are unique marked by cultural values alone; the embodiment of their aesthetics must be redefined in the design process to create both highly functional and emotionally rewarding products.

Author details

I.B. Kashim

Address all correspondence to: ibykash@gmail.com

Department of Industrial Design, Federal University of Technology, Akure, Nigeria

References

[1] Adelabu, O. S. (2011). *Developing a Computer Aided Ceramic Glaze Recipe Using Local Raw Materials in Nigeria*. Unpublished M.Tech Thesis Submitted to the Federal University of Technology, Akure, Nigeria.

[2] Adesanya, A. A. (2010). *Sacred Devotion: Jewellery and Body Adornment in Yoruba culture*. Journal of Critical Interventions. 6.

[3] Ali, R. P. (2007). *Studio Practice in Vocational Education: A Case of Reset Technique in Fabric Decoration*: Multi- Disciplinary Journal of Research Development. April 03, 2012, Great Master Carver Lamidi Fakeye, Retrieved from: http://37thstate.tumblr.com/post/20410112677/great-master-carver-lamidi-fakeye, 8(1), 60-66.

[4] Eicher, J. B. (1976). *Nigerian Handcrafted Textile, Ile Ife*, University of Ife Press.

[5] Emeriewen, K. O. (2007). *A Paradigm for Metal Design*: The University of Benin Art School Experience. Global Journal of Humanities. , 6(152), 51-57.

[6] Hirst, K. (2012). Material Culture. Retrieved from: http://archaeology.about.com/od/mterms/g/material_cultur.htm.

[7] http://wanderingrockswordpress.com/(2009). what-is-visual-culture/

[8] Ibude, I. (2000). *Development of Prototype Solid Fuel Kiln for Cost Effective Ceramics Firing in Nigeria*. Unpublished M.Tech Thesis Submitted to the Federal University of Technology, Akure, Nigeria.

[9] Kashim, I. B. (2004). Sustainability and Economic Survey of Local Ceramic Raw materials for Production of Porcelain Bodies in Nigeria. Unpublished Ph.D. Thesis of the Federal University of Technology, Akure, Nigeria.

[10] Kayode, F. K. (2004). *Analysis of Design Elements in Products of Cane Craft*. Nigerian Crafts and Technology. , 29-39.

[11] Kwanashee, M, Aremu, J. A, Okoi, K, & Oladokun, K. (2000). Research into the Impact of Arts, Culture and Creative Industries on the African Economy; Nigeria. Agoralumier International. , 1-25.

[12] Lam, Y, et al. (2006). Project Chopsticks: *An Asian Life-Style study in Domestic Culinary Habits for Design*. School of Design, the Hong Kong Polytechnic University, Hong Kong.

[13] Maaike, J. (1995). *Influence Upon Sustainable Product Development in the Developing World*. UNEP- WG- SPD Research Programme, Netherland, National Environmental Programme Group. , 1-12.

[14] Makinde, D. O, & Ajiboye, O. J. (2009). *Aso Oke Production and Uses among the Yoruba Southern Nigeria*. The Journal of Pan African Studies. (3) 55- 62.

[15] Moalosi, R, Popovic, V, & Hickling-hudson, A. (2007). *Culture-Oriented Product Design*. International Association of Societies of Design Research.

[16] Ogunduyile, S. R, & Akinbogun, T. L. (2006). *Industrial Design Status and the Challenged to National Development in Nigeria*. Presentation at the SABS Design for Development, Lokgotta, South Africa.

[17] Ojo, E. B. (2000). *Contemporary Printmaking in Nigeria*: An Update; Journal of Creation Arts, Faculty of arts, University of Port Harcourt.

[18] Ojo, E. B. (2004). *Revisiting the Applique and Quilting Craft in Textile Design production*: Nigerian Crafts and Technique, Akure, Nigeria Craft Forum. , 16-20.

[19] Paul, D. (2004). *Visual Culture Isn't First Visual: Multi-literacy, multimodality and meaning*; Studies in Education, A journal of Issues and Research. , 45(3), 252-264.

[20] Salihu, M, Dutsenwai, S. A, & Waziri, M. Y. (2012). *Cultural Industries and wealth Creation: The Case of Traditional Textile Industry,* American International Journal of Contemporary Research. , 2(5), 1-7.

Permissions

The contributors of this book come from diverse backgrounds, making this book a truly international effort. This book will bring forth new frontiers with its revolutionizing research information and detailed analysis of the nascent developments around the world.

We would like to thank Professor Denis A. Coelho, for lending his expertise to make the book truly unique. He has played a crucial role in the development of this book. Without his invaluable contribution this book wouldn't have been possible. He has made vital efforts to compile up to date information on the varied aspects of this subject to make this book a valuable addition to the collection of many professionals and students.

This book was conceptualized with the vision of imparting up-to-date information and advanced data in this field. To ensure the same, a matchless editorial board was set up. Every individual on the board went through rigorous rounds of assessment to prove their worth. After which they invested a large part of their time researching and compiling the most relevant data for our readers. Conferences and sessions were held from time to time between the editorial board and the contributing authors to present the data in the most comprehensible form. The editorial team has worked tirelessly to provide valuable and valid information to help people across the globe.

Every chapter published in this book has been scrutinized by our experts. Their significance has been extensively debated. The topics covered herein carry significant findings which will fuel the growth of the discipline. They may even be implemented as practical applications or may be referred to as a beginning point for another development. Chapters in this book were first published by InTech; hereby published with permission under the Creative Commons Attribution License or equivalent.

The editorial board has been involved in producing this book since its inception. They have spent rigorous hours researching and exploring the diverse topics which have resulted in the successful publishing of this book. They have passed on their knowledge of decades through this book. To expedite this challenging task, the publisher supported the team at every step. A small team of assistant editors was also appointed to further simplify the editing procedure and attain best results for the readers.

Our editorial team has been hand-picked from every corner of the world. Their multi-ethnicity adds dynamic inputs to the discussions which result in innovative outcomes. These outcomes are then further discussed with the researchers and contributors who give their valuable feedback and opinion regarding the same. The feedback is then collaborated with the researches and they are edited in a comprehensive manner to aid the understanding of the subject.

Apart from the editorial board, the designing team has also invested a significant amount of their time in understanding the subject and creating the most relevant covers. They scrutinized every image to scout for the most suitable representation of the subject and create an appropriate cover for the book.

The publishing team has been involved in this book since its early stages. They were actively engaged in every process, be it collecting the data, connecting with the contributors or procuring relevant information. The team has been an ardent support to the editorial, designing and production team. Their endless efforts to recruit the best for this project, has resulted in the accomplishment of this book. They are a veteran in the field of academics and their pool of knowledge is as vast as their experience in printing. Their expertise and guidance has proved useful at every step. Their uncompromising quality standards have made this book an exceptional effort. Their encouragement from time to time has been an inspiration for everyone.

The publisher and the editorial board hope that this book will prove to be a valuable piece of knowledge for researchers, students, practitioners and scholars across the globe.

List of Contributors

Hao Jiang and Ching-Chiuan Yen
Division of Industrial Design, National University of Singapore, Singapore

Carlos A. M. Versos and Denis A. Coelho
Universidade da Beira Interior, Portugal

Helena V. G. Navas
UNIDEMI, Department of Mechanical and Industrial Engineering, Faculty of Science and
Technology, New University of Lisbon, Portugal

Lau Langeveld, René van Egmond, Reinier Jansen and Elif Özcan
Delft University of Technology, The Netherlands

Dian Li, Tom Cassidy and David Bromilow
The School of Design at the University of Leeds, UK

Carlos A.M. Duarte
IADE – Creative University, UNIDCOM - The Research Unit, Lisbon, Portugal

Denis A. Coelho and Sónia A. Fernandes
Universidade da Beira Interior, Portugal

Kristel Dewulf
Howest Industrial Design Center, Kortrijk, Belgium
Delft University of Technology (TUDelft), Faculty of Industrial Design Engineering,
Section, Design for Sustainability, Delft, CE Delft, The Netherlands
Ghent University (UGent), Faculty of Economics and Business Administration,
Department, Management, Innovation and Entrepreneurship, Research Group Corporate
Social Responsibility, Ghent, Belgium

Inalda A. L. L. M. Rodrigues and Denis A. Coelho
Universidade da Beira Interior, Portugal

Chun-Fong You, Chin-Ren Jeng and Kun-Yu Liu
Department of Mechanical Engineering, National Taiwan University, Taiwan, ROC

I.B. Kashim
Department of Industrial Design, Federal University of Technology, Akure, Nigeria

Printed in the USA
CPSIA information can be obtained
at www.ICGtesting.com
JSHW011432221024
72173JS00004B/772